Selected Sensor Circuits

Peter Baumann

Selected Sensor Circuits

From Data Sheet to Simulation

Peter Baumann
Hochschule Bremen
Bremen, Germany

ISBN 978-3-658-38211-7   ISBN 978-3-658-38212-4   (eBook)
https://doi.org/10.1007/978-3-658-38212-4

The translation was done with the help of artificial intelligence (machine translation by the service DeepL.com). A subsequent human revision was done primarily in terms of content.

This Springer imprint is published by the registered company Springer Fachmedien Wiesbaden GmbH, part of Springer Nature.
The registered company address is: Abraham-Lincoln-Str. 46, 65189 Wiesbaden, Germany

Preface to the Third Edition of *Selected Sensor Circuits*

Preface to the Third Edition

When presenting selected sensor circuits, it has proven useful in teaching that first the sensor function is explained on the basis of data sheet information and then a corresponding circuit is analyzed with the program PSPICE.

In the third edition, minor corrections were made to the ultrasonic transmission. The optical sensors are supplemented by an infrared light barrier and the foil force sensors by applications in interaction with a comparator and a triangle-rectangle generator. In the chapter on piezoelectric buzzers, another section on self-triggering buzzers has been added to the existing section on externally triggered buzzers. In the new Chap. 10, surface-wave devices in the form of delay lines, filters, and resonators are presented. This is followed by analyses of a Colpitts and a Pierce oscillator for the frequency of 315 MHz. As in previous years, a sensor circuits project was offered in the summer semester 2019 at the University of Applied Sciences Bremen and successfully completed by a group of students.

I would like to thank Reinhard Dapper, Chief Editor for Electrical Engineering at Springer Vieweg, for his helpful support in the publication of this third extended edition. I would also like to thank Dipl.-Ing. Johannes Aertz, who edited the manuscript according to the publisher's specifications.

Bremen, Germany   Peter Baumann
August 2019

Contents

Temperature Sensors

1.1   NTC Sensors

NTC sensors (Negative Temperature Coefficient) are hot conductors whose resistance does not decrease linearly when heated, because as the temperature rises, more and more charge carriers are released from the metal-oxide ceramic.

Equation (1.1) describes the temperature response of the NTC resistor.

$$R_T = R_N \cdot \exp\left[B \cdot \left(\frac{1}{T} - \frac{1}{T_N}\right)\right] \tag{1.1}$$

Where R_N is the nominal resistance at temperature T_N and B is a material constant in units of Kelvin.

Data Sheet
NTC sensor M87-10, Siemens-Matsushita [1]:
   Glass-encased bead for measuring the temperature of irons or vehicle intake air $R_N = 10$ kΩ at $T_N = 298$ K, $B = 3474$ K.

Task: NTC Sensor Characteristic Curve
For the sensor M87-10 the characteristic curve $R_T = \mathrm{f}(Temp)$ for the temperature range from $-50\,°\mathrm{C}$ to $100\,°\mathrm{C}$ is to be analyzed with the program PSPICE [2].

Solution
In the circuit of Fig. 1.1, Eq. (1.1) is enclosed in curly brackets and entered as the value of the resistor. The parameter specifications were taken from the data sheet.

© The Author(s), under exclusive license to Springer Fachmedien Wiesbaden GmbH, part of Springer Nature 2023
P. Baumann, *Selected Sensor Circuits*,
https://doi.org/10.1007/978-3-658-38212-4_1

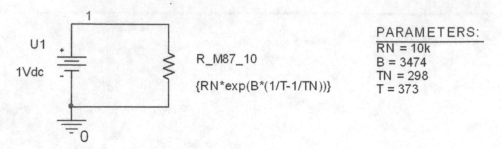

PARAMETERS:
RN = 10k
B = 3474
TN = 298
T = 373

U1

1Vdc

R_M87_10

{RN*exp(B*(1/T-1/TN))}

Fig. 1.1 Circuit for simulating the sensor characteristic curve

Analysis

- PSPICE, Edit Simulation Profile
- Simulation Settings – Fig. 1.1: Analysis
- Analysis type: DC Sweep
- Options: Primary Sweep
- Sweep variable: Global Parameter
- Parameter Name: T
- Sweep type: Linear
- Start value: 223
- End value: 373
- Increment: 1
- Plot, Axis Settings, Axis variable: T-273
- Apply: OK
- PSpice, run

The analysis result according to Fig. 1.2 shows the simulated sensor characteristic curve.

The plot $R_T = f(T)$ with T in Kelvin is available via Plot, Axis Settings, Axis variable, T-273 to obtain the temperature in degrees Celsius on the abscissa.

Task: Temperature Coefficient of the NTC Sensor
Starting from Eq. (1.1), derive the temperature coefficient TK_{RT} of the resistor R_T and plot it with PSPICE for the temperature range from $-50\ °C$ to $100\ °C$.

Solution
Differentiation of Eq. (1.1) gives $dRT/dT = -B - R_T/T^2$. From this follows the temperature coefficient given in Eq. (1.2).

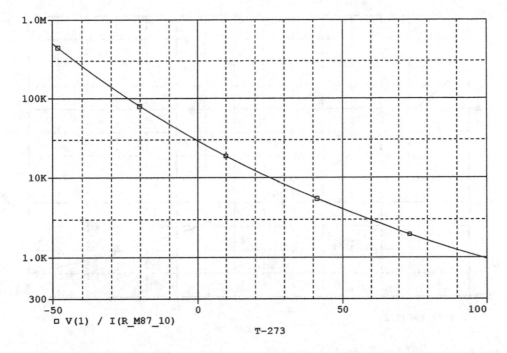

Fig. 1.2 Characteristic curve of the temperature sensor M87_10 with the temperature in °C

$$TK_{RT} = \frac{1}{R_T} \cdot \frac{dR_T}{dT} = -\frac{B}{T^2} \tag{1.2}$$

Analysis
The analysis shall be carried out in the same way as for Fig. 1.2.

The analysis result according to Fig. 1.3 shows that the temperature coefficient does not decrease linearly when the temperature increases.

Task: Temperature Display
Consider the circuit shown in Fig. 1.4. The LED-red is remodeled via a diode Dbreak as follows:

.model LED_red (IS = 1.2E-20 N = 1.46 RS = 2.4 EG = 1.95).

At the temperature of 25 °C, the resistance R-M87-10 has the value of 10 kΩ. Furthermore, $R_4 = 10$ kΩ as well.

- What are the voltages at the inputs of the OP at 25 °C?

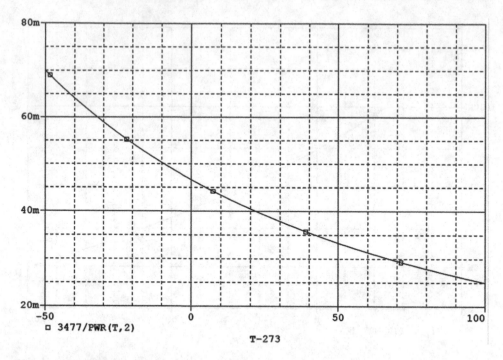

Fig. 1.3 Temperature dependence of the temperature coefficient of R_T

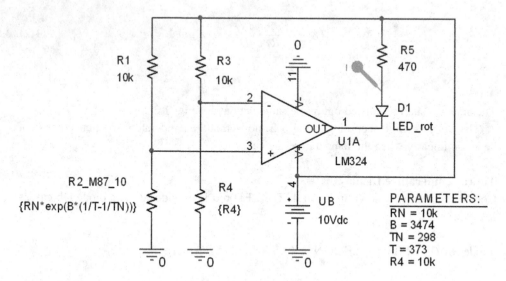

Fig. 1.4 Setting temperatures with the bridge resistor R_4

- Will the LED light up when the temperature rises above 25 °C?
- In principle, what is the effect of changing R_4 to 1.87 kΩ or 104.78 kΩ?

Solution

A voltage of $U_B/2 = 5$ V is applied to each input of the operational amplifier.

   The LED-red lights up at temperatures above 25 °C because $U_N > U_P$.

   It is $I(D_1) = (U_B - U(D_1))/R_5 \approx (10 \text{ V} - 1.6 \text{ V})/470 \ \Omega = 17.8$ mA.

Analysis

- PSpice, Edit Simulation Profile
- Simulation Settings – Fig. 1.4: Analysis
- Analysis type: DC Sweep
- Options: Primary Sweep
- Sweep variable: Global Parameter
- Parameter Name: T
- Sweep type: Linear
- Start value: 233
- End value: 373
- Increment: 0.1
- Options: Parametric Sweep
- Sweep variable: Global Parameter
- Parameter Name: R4
- Sweep type: Value List: 1.87 k, 10 k, 104.78 k
- Plot, Axis Settings, Axis variable: T-273
- Apply: OK
- PSpice, run

The analysis result according to Fig. 1.5 shows:

- with $R_4 = 104.78$ kΩ the LED lights up for *Temp* $\geqq -25$ °C,
- with $R_4 = 10$ kΩ the LED lights up for *Temp* $\geqq 25$ °C and
- with $R_4 = 1.87$ kΩ the LED lights up for *Temp* $\geqq 75$ °C.

1.2   PTC Sensors

PTC sensors (Positive Temperature Coefficient) are PTC thermistors whose resistance increases with increasing temperature because the mobility of the charge carriers decreases.

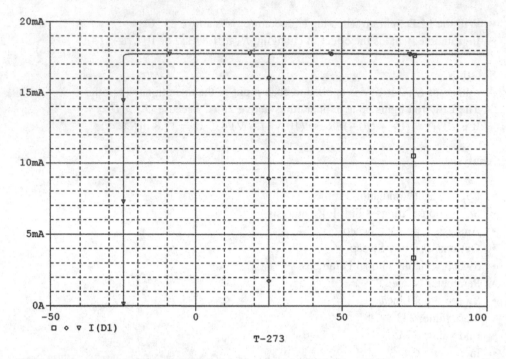

T-273

Fig. 1.5 Depending on R_4, the LED lights up above $-25\,°C$, $25\,°C$ or $75\,°C$

1.2.1   Silicon Temperature Sensors

Silicon temperature sensors are based on the principle of silicon spreading resistance according to the sketch of Fig. 1.6. Starting from an n-silicon single crystal, they are manufactured using planar technology and are used in the temperature range from $-50\,°C$ to $150\,°C$.

The temperature dependence of this sensor type can be described with two temperature coefficients according to Eq. (1.3). For the PTC sensors, the temperature analysis can thus be used in the PSPICE program with the specification in degrees Celsius.

$$R_T = R_{25} \cdot \left[1 + TC_1 \cdot (Temp - Tnom) + TC_2 \cdot (Temp - Tnom)^2 \right] \qquad (1.3)$$

Data Sheet
PTC sensor KTY11-5, Infineon [3]:

$R_{25\,min} = 1950\,\Omega$, $R_{25\,max} = 1990\,\Omega$, it follows that $R_{25} = (R_{25\,min} – R_{25\,max})^{1/2} = 1970\,\Omega$. For the temperature range from $-30\,°C$ to $130\,°C$, Infineon gives the following temperature coefficients:

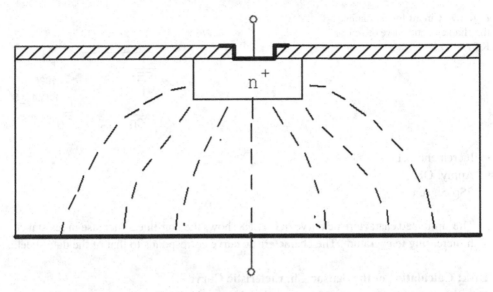

Fig. 1.6 Silicon temperature sensor with current paths

$TC_1 = 7.88\text{--}10^{-3} \text{ K}^{-1}$ and $TC_2 = 1.937\text{--}10^{-5} \text{ K}^{-2}$.

Task: Characteristic curve of the silicon temperature sensor.

With the program PSPICE the characteristic $R_T = f(Temp)$ of the sensor KTY11-5 for the temperature range from $-50\,^{\circ}\text{C}$ to $150\,^{\circ}\text{C}$ is to be analyzed.

Solution

Call a resistor Rbreak from the break library and model it via Edit, PSPICE Model as follows:

.model KTY11_5 RES R = 1 TC1 = 7.88 m, TC2 = 19.37u, TNOM = 25.

The circuit for simulating the characteristic curve is shown in Fig. 1.7.

Analysis

- PSpice, Edit Simulation Profile
- Simulation Settings – Fig. 1.7: Analysis
- Analysis type: DC Sweep
- Options: Primary Sweep
- Sweep variable: Temperature
- Sweep type: Linear
- Start value: −50
- End value: 150

Fig. 1.7 Circuit for simulating
the characteristic curve of sensor
KTY11_5

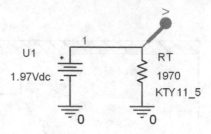

- Increment: 0.1
- Apply: OK
- PSpice, run

The simulated characteristic curve in Fig. 1.8 shows the non-linear increase in resistance with increasing temperature. The characteristic curve corresponds to that of the data sheet.

Task: Calculation of the Sensor Characteristic Curve
A silicon temperature sensor is used with a donor impurity concentration of $N_D = 1.05-10^{14}$ cm^{-3} is produced. At the temperature of 25 °C, the associated electron mobility reaches $\mu_{n25} = 1600$ cm^2/(Vs). The diameter of the contact tip is $d = 30$ μm [4]. The temperature dependence of the electron mobility can be calculated by Eq. (1.4) according to [5]. The temperature T is to be used in Kelvin.

$$\mu_n = \mu_{n25} \cdot \left(\frac{T}{T_{298}}\right)^{-2,42} \tag{1.4}$$

The electrical conductivity of the electrons is described by Eq. (1.5) with

$$\kappa_n = e \cdot N_D \cdot \mu_n \tag{1.5}$$

The sensor resistance is obtained according to Eq. (1.3) via Eq. (1.6).

$$R = \frac{1}{2 \cdot \pi \cdot d \cdot \kappa_n} \tag{1.6}$$

The temperature response of the resistor R in the range from -50 °C to 150 °C is to be determined.

Solution
From Eqs. (1.4), (1.5), and (1.6), the notation of the resistance given by Eq. (1.7) follows to be

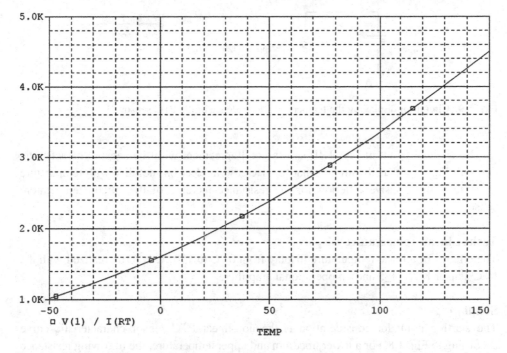

Fig. 1.8 Simulated characteristic curve of the silicon temperature sensor

$$R = R_{25} \cdot \left(\frac{T}{T_{298}}\right)^{2,42} \tag{1.7}$$

with $R_{25} = 1/(2-\pi-d-e-N_D-\mu_{n25}) = 1974 \ \Omega$.

Equation (1.7) can be analyzed with PSPICE. The circuit is shown in Fig. 1.9.

Analysis

- PSpice, Edit Simulation Profile
- Simulation Settings – Fig. 1.9: Analysis
- Analysis type: DC Sweep
- Options: Primary Sweep
- Sweep variable: Global Parameter
- Parameter Name: T
- Sweep type: Linear
- Start value: 233
- End value: 423
- Increment: 10 m
- Plot. Axis Settings, Axis variable: T-273
- Apply: OK
- PSpice, run

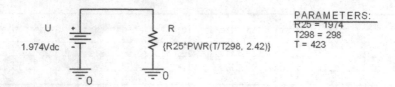

Fig. 1.9 Circuit for displaying the temperature dependence of the PTC sensor

The analysis result in Fig. 1.10 shows good agreement with the sensor characteristic curve in Fig. 1.8, thus illustrating that the decrease in electron mobility with temperature increase is the decisive criterion for the resistance increase of the silicon temperature sensor.

Task: Linear Temperature Display
The circuit for a linear temperature display from 0 °C to 150 °C is to be realized with the PTC sensor KTY11_5 and an operational amplifier.

Solution
The starting point for consideration is the non-linear PTC sensor characteristic curve according to Fig. 1.8. For a lower, medium and upper temperature, the following resistance values are taken from this characteristic curve Eq. (1.6):

$R_{Tu} = 1.6058$ kΩ at 0 °C, $R_{Tm} = 2.8416$ kΩ at 75 °C, and $R_{To} = 4.5067$ at 150 °C.
The next step is to calculate the resistance for linearization according to Eq. (1.8) to the circuit shown in Fig. 1.11.

$$R_{lin} = \frac{R_m \cdot (R_u + R_0) - 2 \cdot R_u \cdot R_0}{R_u + R_0 - 2 \cdot R_m} \qquad (1.8)$$

Using the above resistance values, we get $R_{lin} = 6672$ Ω.
Figure 1.12 shows the linearization achieved.

Analysis

- PSpice, Edit Simulation Profile
- Simulation Settings – Fig. 1.11: Analysis
- Options: Primary Sweep
- Sweep variable: Temperature
- Sweep type: Linear
- Start value: 0
- End value: 150

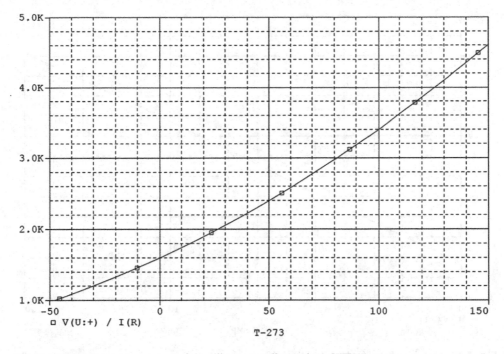

Fig. 1.10 Temperature response of the silicon spreading resistor KTY11

Fig. 1.11 Circuit for linearizing the PTC sensor characteristic curve

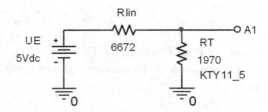

- Increment: 10 m
- Apply: OK
- PSpice, run

At output A_1 the following voltages are obtained for the lower or upper temperature:

$U_u = 0.96992$ V at 0 °C and $U_o = 2.0158$ V at 150 °C.

Given that the output voltage of the operational amplifier should increase linearly from $U_A = 0$ V at the temperature 0 °C to $U_A = 3$ V at the temperature of 150 °C, calculate the voltage gain v_u according to Eq. (1.9).

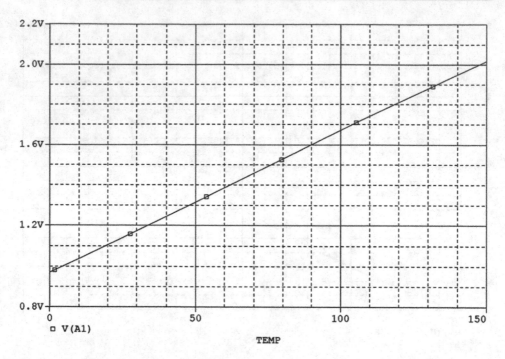

Fig. 1.12 Linearized sensor characteristic curve

$$v_u = \frac{U_{A1o} - U_{A1u}}{U_o - U_u} \tag{1.9}$$

From this it follows that $v_u = (3\text{ V}{-}0\text{ V})/(2.0158\text{ V}{-}0.96992\text{ V}) = 2.8684$.

The resistance R_3 for the circuit shown in Fig. 1.13 follows from Eq. (1.10) given in [6] to be

$$R_3 = v_u \cdot \frac{R_{lin} \cdot R_u}{R_{lin} + R_u} \tag{1.10}$$

We obtain $R_3 = 3713\ \Omega$.

The resistance R_2 can be calculated using Eq. (1.11) according to [6].

$$R_2 = \frac{R_3 \cdot R_u}{R_3 - R_u} \tag{1.11}$$

For this, $R_2 = 2830\ \Omega$.

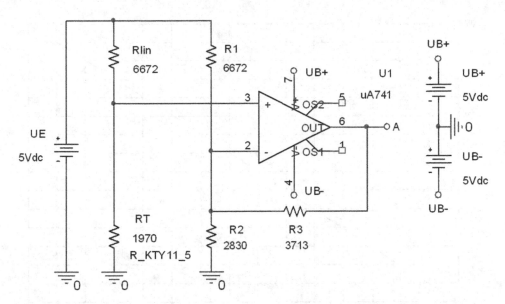

Fig. 1.13 Circuit of the electronic thermometer

Analysis

- PSpice, Edit Simulation Profile
- Simulation settings – Fig. 1.13: Analysis
- Options: Primary Sweep
- Sweep variable: Temperature
- Sweep type: Linear
- Start value: 0
- End value: 150
- Increment: 10 m
- Apply: OK
- PSpice, run

With the diagram according to Fig. 1.14, the objective is fulfilled.

Task: Comparator with PTC Sensor

To be considered is the comparator circuit with the PTC sensor KTY11-5 shown in Fig. 1.15. At 25 °C, $R_T = R_1 = 1970\ \Omega$. The LEDs are modeled as follows [2]:

.model LED_red D (IS = 1.2E-20, N = 1.46, RS = 2.4, EG = 1.95)
. model LED_green D (IS = 9.8E-29, N – 1.12, RS = 24.4, EG = 2.2).

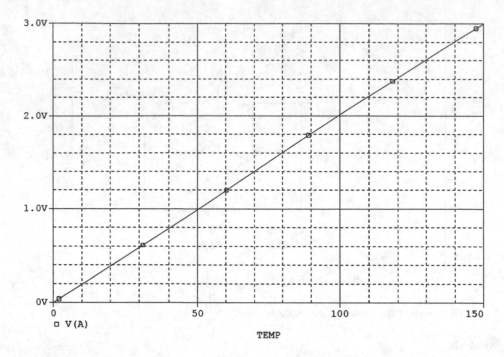

Fig. 1.14 Linear temperature display with the PTC-KTY11-5 sensor thermometer

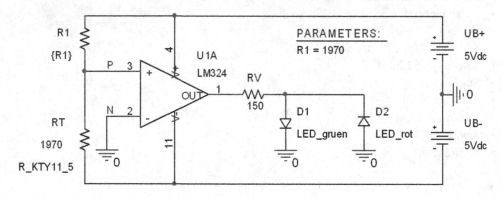

Fig. 1.15 Comparator circuit for displaying temperatures

- Which of the two LEDs lights up when the temperature exceeds 25 °C?
- How does the resistance increase of R_1 affect the LED displays?

Solution

At temperatures above 25 °C, $R_T > R_1$. Approximately U_{B+} is present at the OP output, LED green is lit and LED red is disabled.

For example, if R_1 is increased to the value $R_1 = 3349\ \Omega$, then R_T will reach this value only at the temperature of 100 °C and the LED-green would then light up only above this temperature.

Analysis

- PSpice, Edit Simulation Profile
- Simulation Settings – Fig. 1.15: Analysis
- Options: Primary Sweep
- Sweep variable: Temperature
- Start value: 0
- End value: 150
- Increment: 10 m
- Options: Parametric Sweep
- Sweep variable: Global Parameter
- Parameter Name: R1
- Sweep type: value list: 1970, 3349
- Apply: OK
- PSpice, run

The analysis result according to Fig. 1.16 states:

- with $R_1 = 1970\ \Omega$ lights LED-green at *temp* $\geqq 25$ °C
- with $R_1 = 3349\ \Omega$ lights LED-green at *temp* $\geqq 100$ °C.

For LED-red then applies according to Fig. 1.17:

with $R_1 = 1970\ \Omega$ LED red lights up at *temp* $\leqq 25$ °C.
with $R_1 = 3349\ \Omega$ lights LED-red at *temp* $\leqq 100$ °C.

1.2.2 Platinum Temperature Sensors

At the temperature $T_{nom} = 0$ °C, the Pt 100 sensor exhibits the basic resistance $R_0 = 100\ \Omega$. The sensor can be produced by evaporating a platinum layer onto a ceramic carrier. Because of the relatively low sensor resistance, the lead resistances as well as the self-heating must be taken into account. The positive temperature coefficient is based on the fact that the electrons with higher temperature are slowed down by thermal oscillations.

Data Sheet

In the temperature range from 0 to 850 °C, the following applies to the Pt100 sensor according to Eq. (1.12):

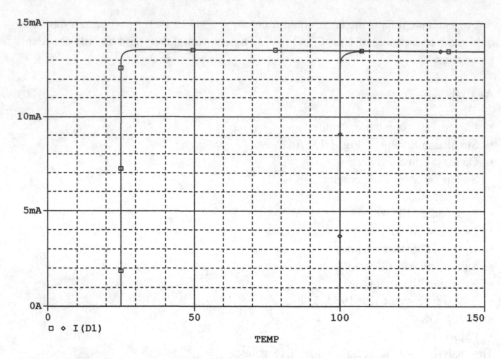

Fig. 1.16 Temperature indication by LED-green

$$R_T = R_0 \cdot \left[1 + TC_1 \cdot (Temp - Tnom) + TC_2 \cdot (Temp - Tnom)^2\right] \qquad (1.12)$$

with the temperature coefficients according to [4]:

$TC_1 = 3.908{-}10^{-3}$ and $TC_2 = -5.802 - 10^{-7}\ \text{K}^{-2}$.
The operating current must be set to $I_B = 1$ mA at $T_{\text{NOM}} = 0\ °C$.

Task: Characteristic Curve of the Pt 100 Sensor
With the circuit shown in Fig. 1.18, the characteristic $R_T = \text{f}(Temp)$ in the range of
$-200\ °C$ to $800\ °C$ to be simulated. The sensor resistance is to be modelled using the
resistance type R_{break} via Edit, PSPICE Model with the above values for TC_1, TC_2 and
T_{NOM} as follows:

.model Pt100 RES R = 1 TC1 = 3908 m TC2 = −0.5082u TNOM = 0.

Analysis

- PSpice, Edit Simulation Profile
- Simulation Settings – Fig. 1.18: Analysis

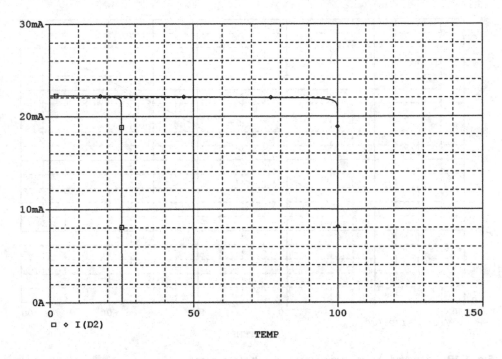

Fig. 1.17 Temperature indication by LED red

Fig. 1.18 Circuit for simulating
the sensor characteristic curve

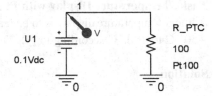

- Analysis type: DC Sweep
- Options: Primary Sweep
- Sweep variable: Temperature
- Sweep type: Linear
- Start value: −200
- End value: 800
- Increment: 0.1
- Apply: OK
- PSpice, run

The analysis result according to Fig. 1.19 shows the braking influence of the temperature coefficient TC_2 on the increase of the PTC sensor.

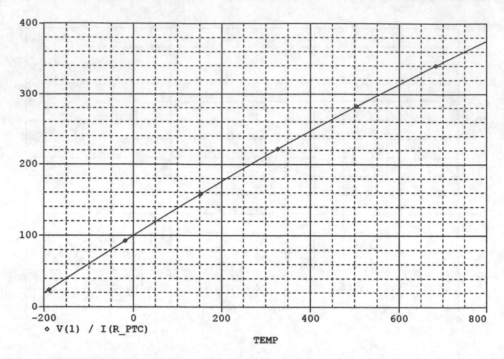

◇ V(1) / I(R_PTC)

TEMP

Fig. 1.19 Simulated characteristic curve of the Pt 100 sensor

Task: Temperature Display with Pt 100 Sensor
A linear temperature display is to be realized with the PTC sensor Pt 100 in the range from
−50 °C to 100 °C according to [7]. The operating current should be $I_B = 1$ mA.

Solution

$U_N = 0.1$ V at 0 °C, $U_N = 0.139$ V at 100 °C,
$I_{R1} = (U_{R4} - U_N)/10$ kΩ = (0.1 V–0.139 V)/10 kΩ = −3.9 µA.
$U_{R2} = I_{R1} \_ R_2 = 30$ µA − 285 kΩ = 1.11 V.
$U_A = U_{R2} + U_N = 1.11$ V + 0.139 V = 1.249 V.
The circuit shown in Fig. 1.20 contains a JFET constant current source and an OP.

Analysis

- PSpice, Edit Simulation Profile
- Simulation Settings – Fig. 1.20: Analysis
- Analysis type: DC Sweep
- Options: Primary Sweep
- Sweep variable: Temperature
- Sweep type: Linear

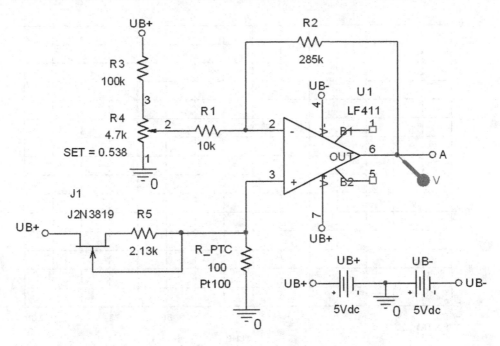

Fig. 1.20 Circuit for temperature display with the Pt 100 sensor

- Start value: −50
- End value: 100
- Increment: 50 m
- Apply, OK
- PSpice, run

As an analysis result according to Fig. 1.21, the output voltage increases in the range of −50 °C to 50 °C with 1 V/100 °C an.

1.3 Bandgap Source as Voltage Reference

With the circuit shown in Fig. 1.22 from [7], an output voltage independent of temperature is achieved, which corresponds approximately to the bandgap voltage of silicon. To achieve the quotient $I_{C1}/I_{C2} \approx 10$, R_2 was fixed at 37 kΩ (instead of 33 kΩ).

The base-emitter voltage of the transistor Q_1 is given by Eq. (1.13) with

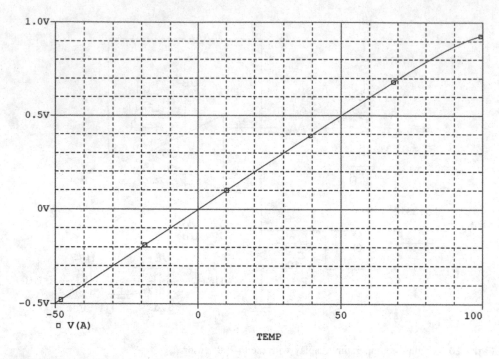

Fig. 1.21 Temperature display for switching with the Pt 100 sensor

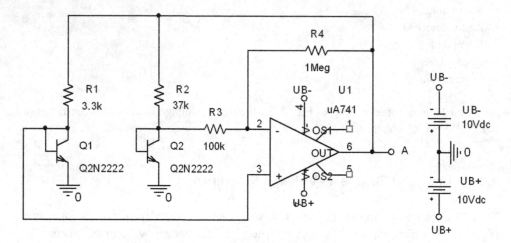

Fig. 1.22 Bandgap source for generating the voltage reference

$$U_{BE1} = U_T \cdot \ln\left(\frac{I_{C1}}{I_S}\right) \tag{1.13}$$

At room temperature, $U_{BE1} \approx 0.6$ V is obtained and for $I_{C1}/I_{C2} = 10$ further $U_{BE2} \approx 0.54$ V. The output voltage follows from Eq. (1.14) to

$$U_A = U_{BE1} - \frac{U_{BE2} - U_{BE1}}{R_3} \cdot R_4 \tag{1.14}$$

One obtains $U_A = 1.2$ V.

For U_{BE1}, the $TK_1 \approx -2$ mV/K and for -$(U_{BE2} - U_{BE1})$- 10 is $TK_2 \approx +2$ mV/K.

Task: Verification of the Temperature-Independent Voltage Reference

In the range from -20 °C to 60 °C, the output voltage curve is to be analysed.

Analysis

- PSpice, Edit Simulation Profile
- Simulation Settings – Fig. 1.22: Analysis
- Analysis type: DC Sweep
- Options: Primary Sweep
- Sweep variable: Temperature
- Sweep type: Linear
- Start value: −20
- End value: 60
- Increment: 0.1
- Apply: OK
- PSpice, run

The analysis result according to Fig. 1.23 shows the reference voltage of approx. 1.28 V, which is almost independent of the temperature, and the base-emitter voltages of the two transistors Q_1 and Q_2. The voltage difference $- 10- (U_{BE2} - U_{BE1})$ or 10-$(U_{BE1} - U_{BE2})$ increases with temperature and provides the temperature compensation to the negative temperature response of U_{BE1}.

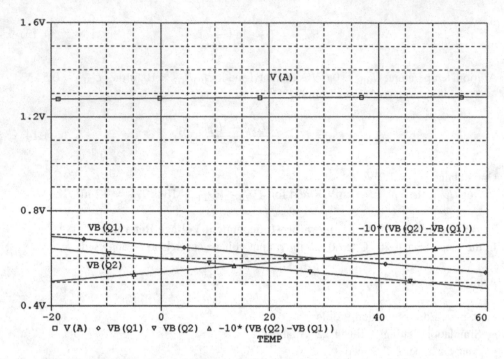

Fig. 1.23 Simulated temperature dependence of stresses of the band gap source

References

1. Siemens-Matsushita: Datenblatt des NTC-Sensors M87-10. München (1998)
2. Baumann, P.: Sensorschaltungen. Vieweg und Teubner, Wiesbaden (2010)
3. Infineon; Datenblatt des PTC-Sensors KTY11-5. München (2000)
4. Elbel, T.: Mikrosensorik. Vieweg, Wiesbaden (1996)
5. Hering, E., Bressler, K., Gutekunst, J.: Elektronik für Ingenieure. VDI, Düsseldorf (1994)
6. Schmidt, W.-D.: Sensor-Schaltungstechnik. Vogel, Würzburg (1997)
7. Kainka, B.: Handbuch der Analogen Elektronik. Franzis, Poing (2000)

Humidity Sensors

2

Humidity is used to measure the water vapour concentration contained in the air. The absolute humidity F_a in the unit g/m^3 corresponds to the quotient of the mass m_w of the water vapor to the volume V_L of the air. The maximum soluble, temperature-dependent humidity is the saturation humidity F_s. From these two quantities, the relative humidity results with $F_r = F_a/F_s$. The relative humidity lies between 0% and 100%.

2.1 Capacitive Humidity Sensors

2.1.1 Sensor KFS 140-D

Figure 2.1 shows the basic structure of a capacitive humidity sensor. Comb-shaped interlocking electrodes embedded in a polymer layer are mounted on a ceramic substrate. The overlying top layer is permeable to moisture [1–3].

At higher humidity, the relative permittivity ε_r of the hygroscopic polymer layer increases, and so does the sensor capacitance according to Eq. (2.1).

$$C = \varepsilon_r \cdot \frac{\varepsilon_0 \cdot A}{d} \tag{2.1}$$

Data Sheet

Capacitive polymer humidity sensor KFS 140-D from Hygrosens [4],
Relative humidity: $F_r = 0$ to 100%, Temperature range: -30–150 °C,
Capacitance: $C = 150$ pF $+/- 50$ pF at $F_r = 30\%$ and 23 °C, Slope: $m = 0.25$ pF/%F$_r$.

© The Author(s), under exclusive license to Springer Fachmedien Wiesbaden GmbH, part of Springer Nature 2023
P. Baumann, *Selected Sensor Circuits*,
https://doi.org/10.1007/978-3-658-38212-4_2

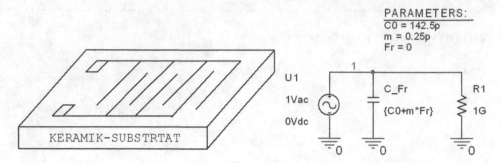

Fig. 2.1 Design and wiring of the humidity sensor KFS140-D

From this follows $C_0 = 144.25$ pF at $F_r = 0\%$ and the sensor capacitance C according to Eq. (2.2) with

$$C = C_0 + m \cdot F_r \tag{2.2}$$

Frequency range: 1–100 kHz, signal form: alternating voltage without DC component. Response time: <12 s. The characteristic $C = f(F_r)$ has a linear increase. Application: air conditioning, ventilation, industrial measurement technology.

Task: Display of the Sensor Characteristic Curve
Using the circuit shown in Fig. 2.1 for the sensor KFS140-D, the dependence of the capacitance C on the relative humidity F_r is to be simulated with the program PSPICE. The range of relative humidity is to be set to $F_r = 0$–100%. The measuring frequency is $f = 10$ kHz.

Analysis

- PSpice, Edit Simulation Profile
- Simulation Settings – Fig. 2.1: Analysis
- Analysis type: AC Sweep/Noise
- AC Sweep type: Logarithmic Decade
- Start Frequency: 10 k
- End Frequency: 10 k
- Points/Decade: 1
- Options: Parametric Sweep
- Sweep variable: Global Parameter
- Parameter Name: Fr
- Sweep type: Linear

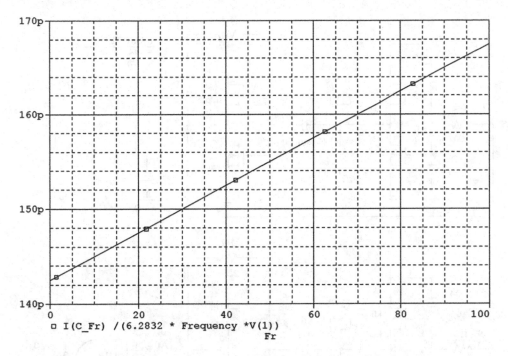

Fig. 2.2 Capacity of the KFS140-D sensor as a function of relative humidity in percent

- Start value: 0
- End value: 100
- Increment: 1
- Apply: OK
- PSpice, run

The analysis result according to Fig. 2.2 shows that the capacitance C *increases* linearly with 0.25 pF/%F_r increases. When the relative humidity is increased from 0% to 100%, the capacitance increases by only about 25 pF.

Task: Astable CMOS Multivibrator
The circuit according to Fig. 2.3 shows two astable multivibrators (AMV) with CMOS inverters. One AMV is determined by the capacitance C_1 of the sensor KFS140-D. The other AMV contains the variable capacitance C_2. The outputs A_1 and A_2 are fed to a NAND gate. The N- and P-channel MOSFETs are to be modeled using MBreakN and MBreakP from the break library. For the protection diode Dbreak has to be called.

. model Mn NMOS W = 100u L = 5u VTO = 1 KP = 30u, KP = UO-ε_0-ε_{ox}/Tox
.model Mp PMOS W = 200u L = 5u VTO = -1 KP = 15u
. model DS D IS = 10f ISR = 1n CJO = 5p.

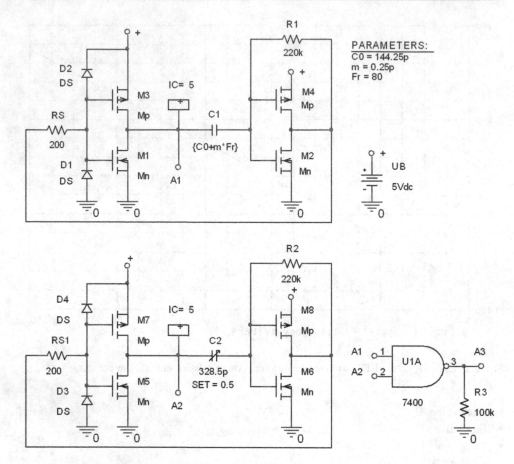

Fig. 2.3 AMV with humidity sensor and comparison AMV

The limiting diodes D_1 and D_2 as well as the resistor R_S belong to the integrated input protection circuit [5]. The pulse frequency f_p is obtained according to [3] with Eq. (2.3) to:

$$f_p = \frac{1}{2 \cdot R1 \cdot C1 \cdot \ln 3} \tag{2.3}$$

The square-wave oscillations of the upper AMV from Fig. 2.3 for the relative humidities $F_r = 0\%$ and 100% are to be shown.

To be used in both cases is an analysis Time Domain (Transient) for the time $\Delta t = 0$ to 250 μs.

Analysis

- PSpice, Edit Simulation Profile
- Simulation Settings – Fig. 2.3: Analysis
- Analysis type: Time Domain (Transient)
- Options: General Settings
- Run to time: 250 us
- Start saving data after: 0
- Transient Options
- Maximum Step size: 0.1 us
- Options: Parametric Sweep
- Sweep variable: Global Parameter
- Parameter Name: Fr
- Sweep type: Value list: 0, 100
- Apply: OK
- PSpice, run

The analysis results according to Figs. 2.4 and 2.5 are similar to the results of Eq. (2.3) with $T = 63.47$ µs at $F_r = 0\%$ and $T = 73.70$ µs at $F_r = 100\%$.

The pulse frequencies follow from this with $f_p = 15.76$ kHz for $F_r = 0\%$ and $f_p = 13.57$ kHz for $F_r = 100\%$.

Task: AMV Comparison

In the circuit shown in Fig. 2.3, the astable multivibrator containing the humidity sensor is set to the parameter value Fr = 80%. The capacitance of the comparator AMV is given the value $C_2 = 164.25$ pF for Fr = 80% by Eq. (2.2). The voltages at the outputs A_1 to A_3 are to be compared.

Analysis

Time Domain (Transient), Run to time: 250 us, Start saving data after: 0, Maximum Step size: 0.1 us, o.k.

The analysis result according to Fig. 2.6 gives identical output voltages U_{A1} and U_{A2} for $C_1 = C_2 = 164.25$ pF. With the dual NAND, negation occurs for U_{A3}.

2.1.2 Sensor KFS 33-LC

Compared to the humidity sensor KFS 140-D, the capacitive polymer humidity sensor KFS 33-LC has a steeper increase of the linear characteristic $C = f(F_r)$. Furthermore, the data sheet contains additional information on the temperature and frequency dependence of this characteristic curve.

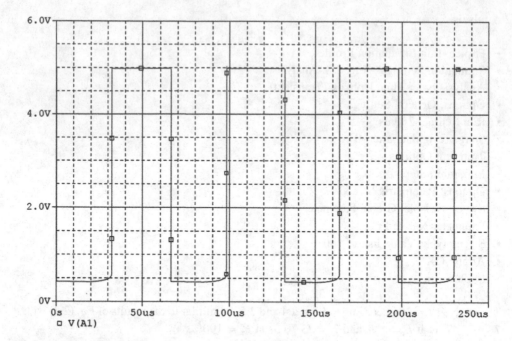

Fig. 2.4 Square-wave oscillations at 0% relative humidity

Data Sheet

- Humidity sensor KFS 33-LC by B + B Thermotechnik [6]
- Relative humidity: $F_r = 0$ to 100%.
- Temperature range: -40–120 °C
- Capacitance: $C = 330$ pF +/− 20 pF at $F_r = 55\%$ and 23 °C
- Slope: $m = 0.6$ pF/%F_r
- Temperature drift: 0.16 pF/K from 5 °C to 70 °C.

For the capacitance, again $C = C_0 + m - F_r$, see Eq. (2.2).

From the values of the manufacturer's characteristic curve $C = 330$ pF at $F_r = 55\%$ and $C = 354$ pF at $F_r = 90\%$ for 23 °C, the slope $m_{23} = 0.68$ pF/%F_r follows. This gives $C_{023} = 292.6$ pF for $F_r = 0\%$.

At $F_r = 55\%$, $C = 330$ pF at 23 °C. With the drift of -0.16 pF/K, it follows that $C = 322.5$ pF at $F_r = 55\%$ and 70 °C. At $F_r = 90\%$ and 70 °C, $C \approx 344$ pF. Thus, the slope $m_{70} = 0.61$ pF/%F_r and $C_{070} = 0.61$ pF for $F_r = 0\%$ at 70 °C. These values are the parameters in the circuits for the simulation.

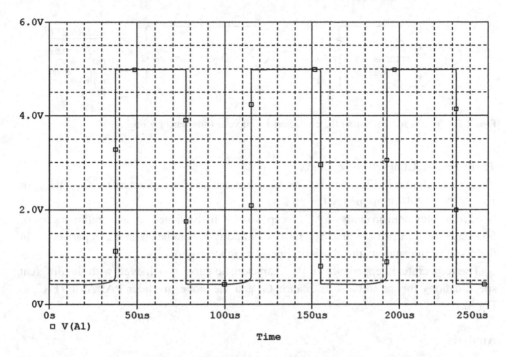

Fig. 2.5 Square-wave oscillations at 100% relative humidity

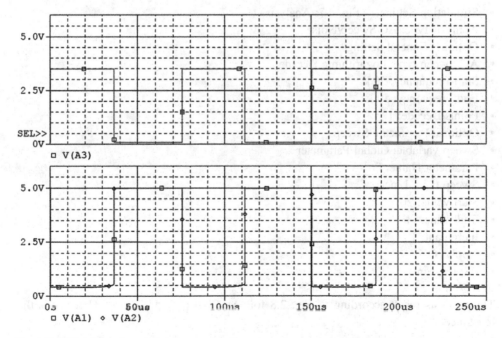

Fig. 2.6 Time dependence of three output voltages

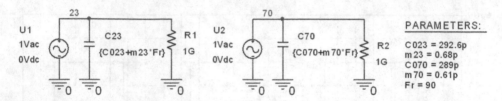

Fig. 2.7 Circuits for simulating the characteristic curves of the sensor KFS 33-LC

Task: Sensor Characteristic Curve

The circuit shown in Fig. 2.7 shall be used to simulate the characteristic $C = f(F_r)$ for $F_r = 20\%$ to 90% at frequency $f = 50$ kHz.

A frequency domain analysis AC Sweep/Noise with the same start and end frequency of 50 kHz is to be carried out first. The relative humidity F_r is to be entered as a parameter. In the result of the analysis the abscissa appears as relative humidity.

The temperature dependence of the characteristic curve is achieved with the different specifications for the initial capacitance C_0 and the slope m under PARAMETERS in Fig. 2.7.

Analysis

- PSpice, Edit Simulation Profile
- Simulation Settings – Fig. 2.7: Analysis
- Analysis type: AC Sweep/Noise
- Options: General Settings
- AC Sweep type: Logarithmic Decade
- Start Frequency: 50 k
- End Frequency: 50 k
- Points/Decade: 1
- Options: Parametric Sweep
- Sweep variable: Global Parameter
- Parameter Name: Fr
- Sweep type: Linear
- Start value: 20
- End value: 90
- Increment: 1
- Apply: OK
- PSpice, run

The analysis result according to Fig. 2.8 follows from $C = 1/X_C - \omega$. Here X_C is the reactance.

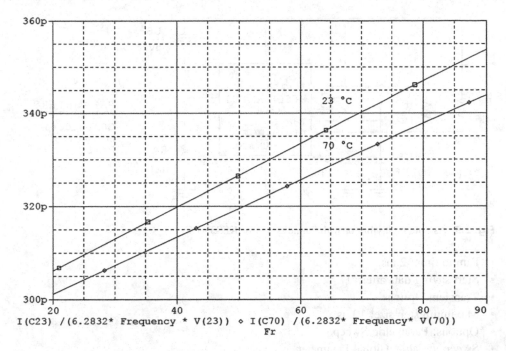

Fig. 2.8 Dependence of sensor capacity on relative humidity in percent

The characteristic curves show the linear increase of the sensor capacity with increasing humidity. The simulated temperature dependence of the sensor characteristics is reproduced as in the data sheet.

Task: Astable 555D Multivibrator

The humidity sensor KFS33-LC is to be used as capacitance C_S in the astable multivibrator with the timer circuit 555D according to Fig. 2.9.

The pulse frequencies f_p according to Eq. (2.4) for the relative humidity values $F_r = 20\%$ and $F_r = 90\%$ are to be compared.

$$f_p = \frac{1.44}{(R_1 + 2 \cdot R_2) \cdot C_S} \tag{2.4}$$

Analysis

- PSpice, Edit Simulation Profile
- Simulation Settings – Fig. 2.9: Analysis
- Analysis type: Time Domain (Transient)
- Options: General Settings

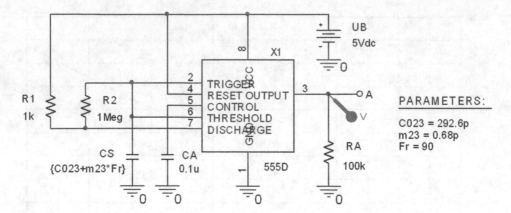

Fig. 2.9 Astable multivibrator with humidity sensor KFS33-LC

- Run to time: 2 ms
- Start saving data after: 0
- Transient Options
- Maximum step size: 10 us
- Options: Parametric Sweep
- Sweep variable: Global Parameter
- Parameter Name: Fr
- Sweep type: Value list: 20, 90
- Apply: OK
- PSpice, run

Equation (2.4) gives: $f_p = 2350$ kHz at $F_r = 20\%$ and $f_p = 2034$ kHz at $F_r = 90\%$. The corresponding values of period $T = 425.49$ μs for $F_r = 20\%$ and $T = 491.63$ μs for $F_r = 90\%$ are approximately confirmed with Figs. 2.10 and 2.11. The difference in pulse frequencies for $F_r = 20\%$ and 90% is relatively small. An astable multivibrator with fixed capacitors can be used to generate differential pulses.

Task: Capacitance Measuring Bridge

Figure 2.12 shows a capacitance measuring bridge according to [7]. The capacitance C_1 with the given parameters corresponds to the humidity sensor KFS33-LC at the temperature of 23 °C. The three remaining capacitances are set to the nominal sensor capacitance $C = 330$ pF, which is valid for relative humidity $F_r = 55\%$. The diagonal voltage $U_d = U_A - U_B$ for the values $F_r = 40\%$, 55% and 90% is to be plotted. Expectation: $U_d = 0$ for $F_r = 55\%$.

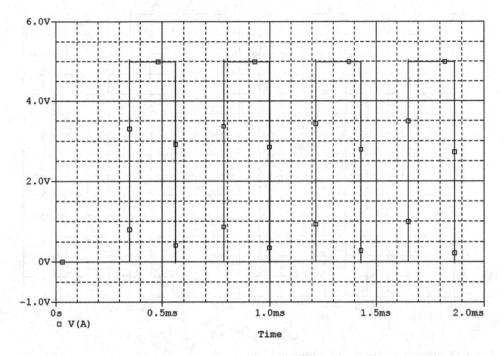

Fig. 2.10 Vibration curve at 20% relative humidity

Analysis

- PSpice, Edit Simulation Profile
- Simulation Settings – Fig. 2.12: Analysis
- Analysis type: Time Domain (Transient)
- Options: General Settings
- Run to time: 80 us
- Start saving data after: 0
- Transient Options
- Maximum step size: 0.1 us
- Options: Parametric Sweep
- Sweep variable: Global Parameter
- Parameter Name: Fr
- Sweep type: Value list: 40, 55, 90
- Apply: OK
- PSpice, run

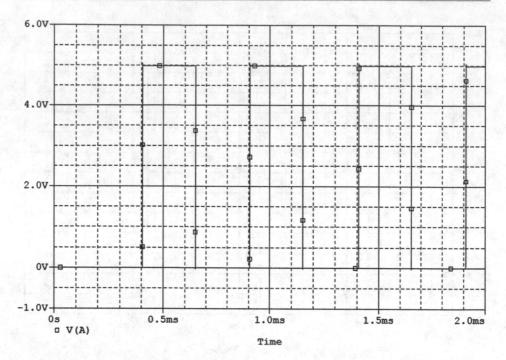

Fig. 2.11 Vibration curve at 90% relative humidity

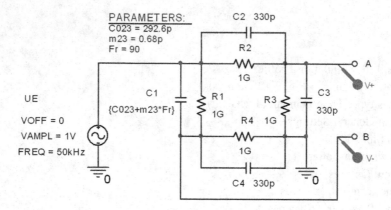

Fig. 2.12 Capacitance measuring bridge with humidity sensor KFS33-LC

The analysis result shown in Fig. 2.13 shows the diagonal voltage of zero volts for $F_r = 55\%$ with the bridge tuned. For $F_r = 90\%$ the sine peak is higher than for $F_r = 40\%$.

Between these two values, which are respectively above and below the standard of $F_r = 55\%$, a phase shift of $180°$ occurs.

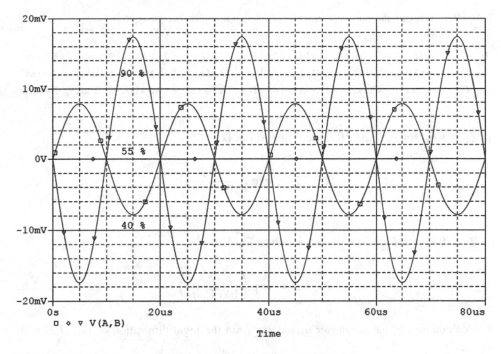

Fig. 2.13 Diagonal stresses for three values of relative humidity

2.2   Electrolytic Humidity Sensor EFS-10

The electrolytic humidity sensor EFS-10 shows a strong reduction of its impedance with increasing relative humidity.

Data Sheet
Electrolytic polymer sensor from B + B Thermotechnik [6].

- Relative humidity: $F_r = 20\text{–}95\%$,
- Temperature range: 0–60 °C
- Signal form: AC voltage with 1 V_{eff}, Measuring frequency: 1 kHz (0.1–5 kHz)
- Application: climate control, building services, air dryer

Table 2.1 shows the dependence of the impedance Z on the relative humidity F_r at three temperatures as an excerpt from the manufacturer's specifications.

The characteristic $\log(Z) = f(F_r)$ consists of two sections with different slopes at the respective temperatures:

Table 2.1 Sensor impedance as a function of relative humidity according to [6]

	20%	30%	40%	50%	60%	70%	80%	90%
10 °C	9 MΩ	2.5 MΩ	740 kΩ	220 k	72 kΩ	25.8 kΩ	9.50 kΩ	3.72 kΩ
25 °C	2.89 MΩ	900 kΩ	270 kΩ	81 kΩ	33 kΩ	13 kΩ	5.30 kΩ	2.20 kΩ
40 °C	1.3 MΩ	420 kΩ	135 kΩ	45 kΩ	18.20 kΩ	7.40 kΩ	3.22 kΩ	1.41 kΩ

1. $Z_1 = f(F_r)$ for $F_r = 20\text{--}50\%$ according to Eq. (2.5)

$$Z_1 = Z_{01} \cdot \exp\left(\frac{-F_r}{100\% \cdot n_1}\right) \qquad (2.5)$$

2. $Z_2 = f(F_r)$ for $F_r = 50\text{--}90\%$ according to Eq. (2.6)

$$Z_2 = Z_{02} \cdot \exp\left(\frac{-F_r}{100\% \cdot n_2}\right) \qquad (2.6)$$

The calculation of the parameter n_1 follows from the logarithmization of Eq. (2.5) with Eq. (2.7)

$$n_1 = \frac{F_{rb} - F_{ra}}{100\% \cdot \ln\left(Z_{1a}/Z_{1b}\right)} \qquad (2.7)$$

At the temperature of **25 °C** one obtains with the values $Z_{1a} = 2890$ kΩ at $F_{ra} = 20\%$ and $Z_{1b} = 81$ kΩ at $F_{rb} = 50\%$ from Table 2.1 the result: $\boldsymbol{n_1 = 84\text{--}10^{-3}}$. From Eq. (2.5) it follows as a further parameter: $\boldsymbol{Z_{01} = 31.32\text{--}10^6\ \Omega}$.

Similarly, from Eq. (2.6), one arrives at the parameter n_2 according to Eq. (2.8).

$$n_2 = \frac{F_{rd} - F_{rc}}{100\% \cdot \ln\left(Z_{1c}/Z_{1d}\right)} \qquad (2.8)$$

At the temperature of **25 °C** one calculates with the values $Z_{2c} = 81$ kΩ at $F_{rc} = 50\%$ and $Z_{2d} = 2.2$ kΩ at Fr = 90% the parameters: $\boldsymbol{n_2 = 111\text{--}10^{-3}}$ and $\boldsymbol{Z_{02} = 7.35\text{--}10^6\ \Omega}$.

At the temperatures of 10 °C and 40 °C, respectively, the characteristic curves according to the data sheet in both humidity sections run largely parallel to those with the temperature of 25 °C, so that the values of n_1 and n_2 determined there can be adopted. For the temperature of 10 °C the following parameters are calculated: $\boldsymbol{n_1 = 84\text{--}10^{-3}}$ and $\boldsymbol{Z_{01} = 97.34\text{--}10^6\ \Omega}$ as well as $\boldsymbol{n_2 = 111\text{--}10^{-3}}$ and $\boldsymbol{Z_{02} = 19.89\text{--}10^6\ \Omega}$. For the temperature of 40 °C the parameters follow: $\boldsymbol{n_1 = 84\text{--}10^{-3}}$ and $\boldsymbol{Z_{01} = 14.06\text{--}10^6\ \Omega}$ as well as $\boldsymbol{n_2 = 111\text{--}10^{-3}}$ and $\boldsymbol{Z_{02} = 4.07\text{--}10^6\ \Omega}$.

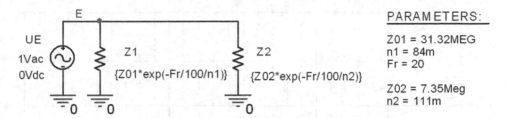

Fig. 2.14 Circuit for simulating the characteristic curves for two humidity ranges

Task: Characteristics of the Electrolytic Humidity Sensor

Using the circuit shown in Fig. 2.14, the characteristic curves $Z_1 = f(F_r)$ in the range $F_r = 20–50\%$ and $Z_2 = f(F_r)$ in the range $F_r = 50–90\%$ shall be simulated. The analysis shall be performed for the temperature of 25 °C and extended to temperatures of 10 and 40 °C.

Analysis

- PSpice, Edit Simulation Profile
- Simulation Settings – Fig. 2.14: Analysis
- Analysis type: AC Sweep/Noise
- AC Sweep type: Logarithmic, Decade
- Start Frequency: 1 kHz
- End Frequency: 1 kHz
- Points/Decade: 1
- Options: Parametric Sweep
- Sweep variable: Global Parameter
- Parameter Name: Fr
- Sweep type: Linear
- Start value: 20
- End value: 50
- Increment: 10 m
- Apply: OK
- PSpice, run

The analysis result according to Fig. 2.15 shows the exponential decrease of the sensor resistance with an increase of the relative humidity. At higher temperatures a further decrease of the resistance occurs. The values of Table 2.1 are fulfilled.

The following analysis applies to the higher relative humidity range.

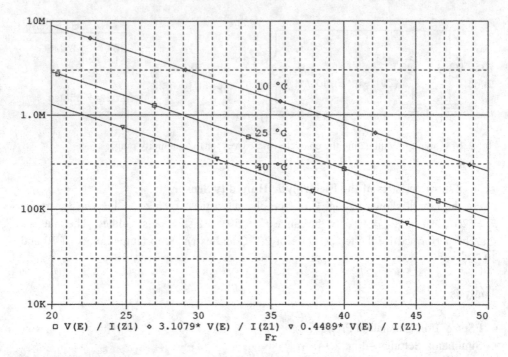

Fig. 2.15 Characteristics of the EFS-10 sensor in the low humidity range

Analysis

- PSpice, Edit Simulation Profile
- Simulation Settings – Fig. 2.14: Analysis
- Analysis type: AC Sweep/Noise
- AC Sweep type: Logarithmic, Decade
- Start Frequency: 1 kHz
- End Frequency: 1 kHz
- Points/Decade: 1
- Options: Parametric Sweep
- Sweep variable: Global Parameter
- Parameter Name: Fr
- Sweep type: Linear
- Start value: 50
- End value: 90
- Increment: 10 m
- Apply: OK
- PSpice, run

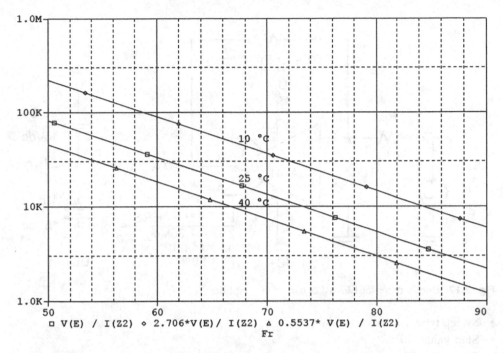

Fig. 2.16 Characteristics of the EFS-10 sensor in the high humidity range

The simulated characteristic curves according to Fig. 2.16 show a lower slope than the characteristic curves shown in Fig. 2.15.

Task: Amplifier with Electrolytic Humidity Sensor

In the circuit shown in Fig. 2.17, the electrolytic humidity sensor EFS-10 is arranged in the feedback branch of the inverting amplifier. The output voltage is to be analysed as a function of the relative humidity in the range $F_r = 50\text{–}90\%$.

Analysis

- PSpice, Edit Simulation Profile
- Simulation Settings – Fig. 2.17: Analysis
- Analysis type: AC Sweep/Noise
- AC Sweep type: Logarithmic, Decade
- Start Frequency: 1 kHz
- End Frequency: 1 kHz
- Points/Decade: 1
- Options: Parametric Sweep
- Sweep variable: Global Parameter
- Parameter Name: Fr

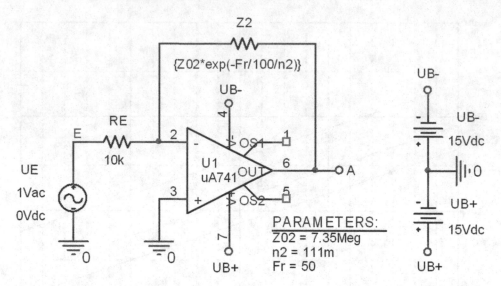

Fig. 2.17 Humidity sensor EFS-10 as part of the amplifier

- Sweep type: Linear
- Start value: 50
- End value: 90
- Increment: 10 m
- Apply: OK
- PSpice, run

Figure 2.18 shows the output voltage and the natural logarithm of this voltage.

2.3   Resistive Humidity Sensor

The circuit shown in Fig. 2.19 contains a comb-like electrode arrangement on a sponge-like base [5]. When wetted with moisture, the resistance of this arrangement drops sharply. The modelling of the LED corresponds to that of Fig. 1.4.

Task: Warning Signal for Humidification
It is to be analysed whether the LED lights up after 5 s, i.e. after humidification has started.

Analysis

- PSpice Simulation Profile
- Simulation Settings – Fig. 2.19: Analysis
- Analysis type: Time domain (Transient)

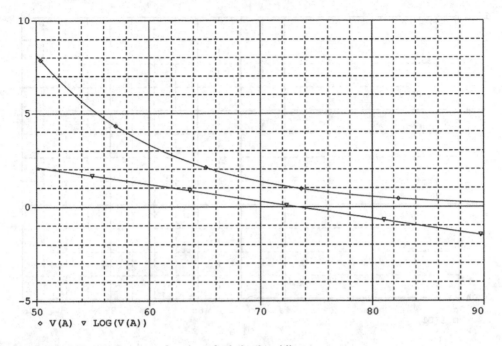

◇ V(A) ▽ LOG(V(A))

Fig. 2.18 Output voltage as a function of relative humidity

Fig. 2.19 Circuit with resistive
humidity sensor

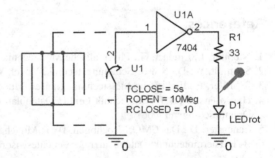

- Options: General settings
- Run to time. 10 s
- Start saving data after: 0 s
- Maximum step size: 10 ms
- Apply: OK
- PSpice, run

When wet, the electrode arrangement becomes conductive. The input voltage U_E goes from HIGH to LOW after the specified 5 s and the output voltage U_A goes from LOW to HIGH, see Fig. 2.20.

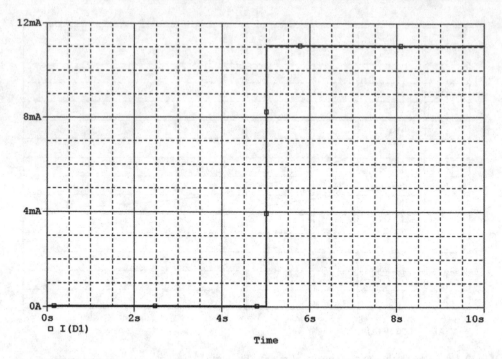

Fig. 2.20 LED display when moisture occurs

References

1. Niebuhr, J., Lindner, G.: Physikalische Messtechnik mit Sensoren. Oldenbourg, München (2011)
2. Schmidt, W.-D.: Sensor-Schaltungstechnik. Vogel, Würzburg (1997)
3. Schrüfer, E.: Elektrische Messtechnik. Hanser, München (2001)
4. Hygrosens: B+B Thermotechnik GmbH Datenblatt des kapazitiven Feuchtesensors KFS 140-D, Donaueschingen (2006)
5. Lancaster, D.: Das CMOS-Kochbuch. IWT, München (1994)
6. B+B Thermotechnik: Datenblätter der Feuchtesensoren KFS 33-LC und EFS-10, Donaueschingen (2013)
7. Wirsum, S.: Das Sensor-Kochbuch. IWT, Bonn (1994)

Optical Sensors

3

3.1 Photoresistors

Photoconductors consist of cadmium sulfide for wavelengths of 0.5–0.7 μm or of lead sulfide for the infrared range with 1–3 μm. With increasing exposure to light, charge carriers are increasingly released from the crystal lattice, causing the electrical resistance to drop. Photoresistors react sluggishly to changes in light. Their application is in twilight switches, light barriers, light meters, optocouplers and dimmers. The maximum eye sensitivity is 0.555 μm.

Data Sheet

Sensor A 1060-12, Transfer Multisort Elektronik, Lodz [1].

R_{10} = **18.4 kΩ** at E_v = 10 lx, R_{100} = 3.8 kΩ at E_v = 100 lx, U_{max} = 200 V, P_{max} = 75 mW.

The decrease in resistance with increasing illuminance E_v describes Eq. (3.1)

$$\frac{R_{10}}{R_{100}} = \left(\frac{E_{v100}}{E_{v10}}\right)^{\gamma} \tag{3.1}$$

From this follows the exponent γ with Eq. (3.2).

$$\gamma = \frac{\lg(R_{10}/R_{100})}{\lg(E_{v100}/E_{v10})} \tag{3.2}$$

For the sensor A 1060_12, we obtain $\gamma_{10/100}$ = **0.685**. In the range E_v = 10 lx to 1 klx, Eq. (3.3) applies approximately to the photoresistance R_p.

P. Baumann, *Selected Sensor Circuits*,
https://doi.org/10.1007/978-3-658-38212-4_3

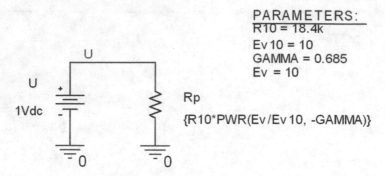

PARAMETERS:
R10 = 18.4k
Ev 10 = 10
GAMMA = 0.685
Ev = 10

Fig. 3.1 Simulation of the dependence of the photoconductance on the illumination intensity

$$R_{p \approx R_{10}} \cdot \left(\frac{E_v}{E_{v10}} \right)^{-\gamma} \tag{3.3}$$

Task: Representation of the Characteristic Curve of the Photoresistor
The circuit shown in Fig. 3.1 is used to represent the characteristic curve $R_p = f(E_v)$ of the sensor A1060-12 for the range of illuminance $E_v = 10$ to 300 lx.

Analysis

- PSpice, Edit Simulation Profile
- Simulation settings – Fig. 3.1: Analysis
- Analysis type: DC Sweep
- Options: Primary Sweep
- Sweep variable: Global Parameter
- Parameter Name: Ev
- Sweep type: Logarithmic Decade
- Start value: 10
- End value: 300
- Points/Decade: 100
- Apply: OK
- PSpice, run

With the analysis result according to Fig. 3.2, the specifications of the data sheet are fulfilled.

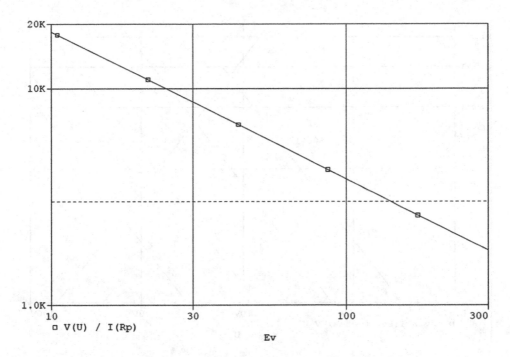

Fig. 3.2 Simulated characteristic curve of the photoresistor A 1060-12 with Ev in Lux

Task: Current-Voltage Characteristics of the Photoresistor

The *I-V characteristics of* the photoresistor A1060-12 for $U = 0$ V to 50 V with the illuminance $E_v = (10, 30, 100, 300)$ lx as parameter and the maximum power dissipation $P_{max} = 75$ mW are to be simulated, see the circuit according to Fig. 3.1.

Analysis

- PSpice, Edit Simulation Profile
- Simulation Settings – Fig. 3.1
- Analysis type: DC Sweep
- Options: Primary Sweep
- Sweep variable: Voltage Source
- Name: V_U
- Sweep type: Logarithmic Decade
- Start value: 1 V
- End value: 100 V
- Points/Decade: 100
- Options: Parametric sweep
- Sweep variable: Global Parameter

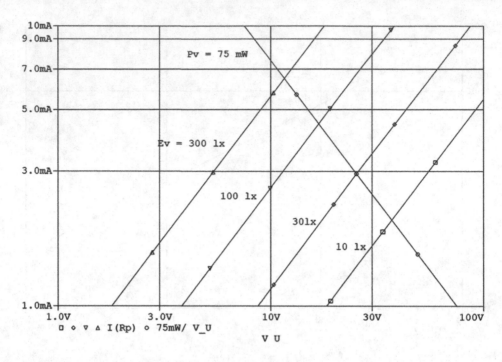

Fig. 3.3 Current-voltage characteristics of the photoresistor

- Parameter Name: Ev
- Sweep type: Value list: 10, 30, 100, 300
- Takeover: OK
- PSpice, run

The analysis result according to Fig. 3.3 shows that the photoresistance $R_p = \text{V\_U}/\text{I}(R_p)$ of the sensor A1060-12 decreases with increasing illumination. The power dissipation hyperbola for $P_v = 75$ mW appears as a straight line in the log-log scale.

Task: Controlling a Power MOSFET with the Photoresistor
In the circuit shown in Fig. 3.4, a power MOSFET is driven by a photoresistor and feeds a lamp [2]. It is to be simulated how the lamp current depends on the illuminance E_v.

Analysis

- PSpice, Edit Simulation Profile
- Simulation Settings – Fig. 3.4: Analysis
- Analysis type: DC Sweep
- Sweep variable: Global Parameter
- Parameter Name: Ev

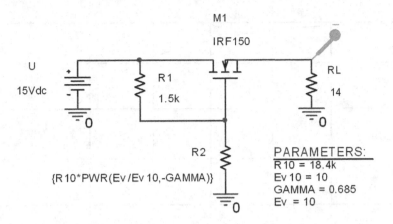

Fig. 3.4 Controlling a lamp via the A 1060-12 photoresistor

- Sweep type: Logarithmic Decade
- Start value: 1
- End value: 4 k
- Points/Decade: 100
- Takeover: OK
- PSpice, run

The analysis result according to Fig. 3.5 shows that the lamp current is reduced with increasing illuminance E_v due to the lowering of the gate potential at the MOSFET.

3.2 Light Control of a DC Micromotor

For this section, the deflection of a current conductor in the magnetic field is first considered in the representation according to Fig. 3.6.

If a current-carrying, straight conductor with the magnetic field surrounding it enters the relevant field of a permanent magnet, then it is deflected laterally by a force F, see Eq. (3.4).

$$F = l \cdot B \cdot I \tag{3.4}$$

The conductor length, B *is* the magnetic induction (flux density) and I is the current. The following applies for the dimension: $F = [\text{m-Vs/m}^2 \text{ A}] = [\text{Ws/m}] = [\text{N}]$. Furthermore, $1 \text{ Ws} = 1 \text{ N m}$.

The torque M according to Eq. (3.5) is determined by the force F and the radius r *of* the rotor according to Eq. (3.5). Here the force follows from the interaction of the stator magnetic field with the current in the rotor bars of the rotor winding, see Eq. (3.4).

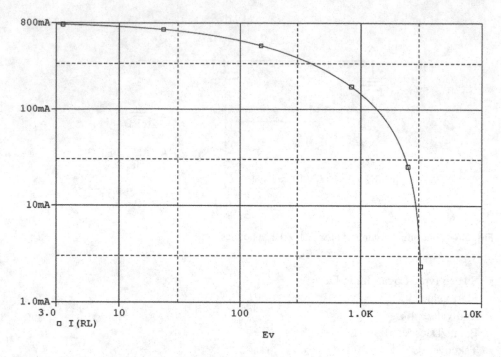

Fig. 3.5 Simulated dependence of lamp current on illuminance

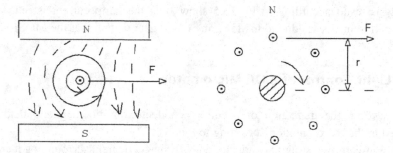

Fig. 3.6 Force effect on current-carrying conductors in the magnetic field

$$M = F \cdot r \tag{3.5}$$

The output power P_2 is obtained from the product of the torque M and the angular velocity ω using Eq. (3.6).

$$P_2 = M \cdot \omega \tag{3.6}$$

Table 3.1 Characteristics of the DC micromotor 018 S for $M = 3$ mN-m from Faulhaber [3]

Rated voltage	$U_N = 18$ V	No-load current constant	$k_0 = 0.99$
Terminal resistance	$R = 25$ Ω	Torque constant	$k_M = 19.6$ mN-m/A
Terminal inductance	$L = 600$ μH	Generator voltage constant	$k_E = 2.05$ mV/rpm
No-load current	$I_0 = 7$ mA	Rotor moment of inertia	$J = 2.1$ gcm^2

The relationship between the angular velocity ω and the rotational speed n is described by Eq. (3.7).

$$\omega = 2 \cdot \pi \cdot n \qquad (3.7)$$

The type of motor used is a DC micromotor whose rotor is designed as a cantilevered copper coil. This bell-shaped, ironless rotor rotates around a cylindrical core magnet. The DC micromotor is characterised by a very low moment of inertia and is cogging-free.

Data Sheet
Table 3.1 lists the main characteristics of the motor used.

Development of the Engine SPICE Model
The total current of the motor according to Eq. (3.8) is the sum of the no-load current I_0 and the armature current I_A, which is independent of the applied voltage U.

$$I = \frac{U \cdot (1 - k_0)}{R} + \frac{M}{k_M} \qquad (3.8)$$

The no-load current constant k_0 follows from Eq. (3.9) with

$$k_0 = 1 - \frac{I_0 \cdot R}{U_N} \qquad (3.9)$$

The torque constant k_M corresponds to the quotient of the motor torque M to the absorbed current.

Finally, the voltage constant k_E is used to record the quotient of the voltage U_i induced in the armature to the angular velocity ω. Furthermore, the relationship to the motor constant k_M is established, see Eq. (3.10). Thereby $\omega = n \cdot 2 \cdot \pi/60$, with the speed n in rpm, that is, revolutions per minute.

$$k_E = \frac{U_i}{\omega} = \frac{2 \cdot \pi \cdot k_M}{60} \qquad (3.10)$$

The speed n can be calculated using Eq. (3.11) as follows:

$$n = \frac{U - I \cdot R}{k_E} \tag{3.11}$$

The starting inertia of the motor is determined by the inductance L and the moment of inertia J. The deceleration occurring during the current rise in the coil results from the electrical time constant τ_e according to Eq. (3.12).

$$\tau_e = \frac{L}{R} \tag{3.12}$$

The mechanical start-up time constant τ_m corresponds to the time that elapses before the motor operated without a load reaches 63% of its speed, see Eq. (3.13) according to [3].

$$\tau_m = \frac{R \cdot J}{k_M{}^2} \tag{3.13}$$

In the circuit shown in Fig. 3.7, the GPOLY type voltage controlled current source GI senses the current I according to Eq. (3.8), while the source G_n can be used to represent the speed n according to Eq. (3.11) [4].

With the capacity C according to Eq. (3.14), the starting inertia [5] can be taken into account.

$$C = \frac{J}{(k_M)^2} \tag{3.14}$$

The specified rotor moment of inertia $J = 2.1 \text{ gcm}^2$ corresponds to $J = 210\text{--}10^{-9} \text{ Ws}^3$.

The power input P_1 is obtained according to Eq. (3.15) as follows

$$P_1 = U \cdot I \tag{3.15}$$

The output power P_2 follows from Eq. (3.16). The torque M must be entered in Nm and the speed n in rpm.

$$P_2 = \frac{2 \cdot \pi}{60} \cdot M \cdot n \tag{3.16}$$

The efficiency η in percent is given by Eq. (3.17) with

$$\eta = 100 \cdot \frac{P_2}{P_1} \tag{3.17}$$

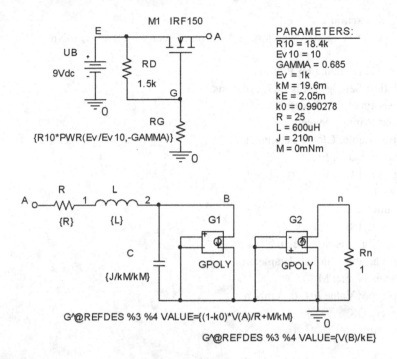

Fig. 3.7 Light control of the DC micromotor type 018 S

Task: Light Control

Light control of the motor via a photoresistor in conjunction with the N-channel enhancement MOSFET IRF 150 is to be carried out, see Fig. 3.7.

Analysis for Variant 1

- PSpice, Edit Simulation Profile
- Simulation Settings – Fig. 3.7: Analysis
- Analysis type: DC Sweep
- Options: Primary Sweep
- Sweep variable: Global Parameter
- Parameter Name: Ev
- Sweep type: Linear
- Start value: 1 lx
- End value: 1 klx
- Increment: 1 lx
- Takeover: OK
- PSpice, run

Analysis of Variant 2
With U_B from 9 V to 20 V and torque M as parameter with 0 and 3 mNm.

- PSpice, Edit Simulation Profile
- Simulation Settings – Fig. 3.7: Analysis
- Analysis type: DC Sweep
- Options: Primary Sweep
- Sweep variable: Global Parameter
- Parameter Name: Ev
- Sweep type: Linear
- Start value: 1 lx
- End value: 1 klx
- Increment: 1 lx
- Options: Parametric Sweep
- Sweep variable: Global Parameter
- Parameter Name: M
- Sweep type: Value list: 0, 3 mNm
- Takeover: OK
- PSpice, run

The dependence of the speed n on the illuminance E_v in the range $E_v = 0$ to 1000 lx at different values of the operating voltage and the torque is to be investigated.

The following settings must be made in detail:

Variant 1: $n = f(E_v)$ at $U_B = 9$ V and $M = 0$ and
Variant 2: $n = f(E_v)$ at $U_B = 20$ V and $M = 0$ mNm as well as $M = 3$ mNm.

The analysis result according to Fig. 3.8 shows that the speed decreases with increasing illuminance E_v because the potential at nodes G and A decreases. Compared to the voltage $U = V(A)$, the product I_0–R in Eq. (3.11) is relatively small.

The analysis result according to Fig. 3.9 shows that the speed n decreases sharply when the motor is loaded with $M = 3$ mN-m, especially because the current $I = I(R)$ increases sharply, see Eq. (3.11).

3.3   Silicon Photodiode

The mode of operation of the pn photodiode is based on the fact that the electron-hole pairs generated in the junction layer with the incident light quanta are separated by the electric field strength. In the pin photodiode, this process occurs mainly in the middle intrinsic layer. The electrons drift to the positive pole and the holes to the negative pole of the applied reverse voltage, see Fig. 3.10.

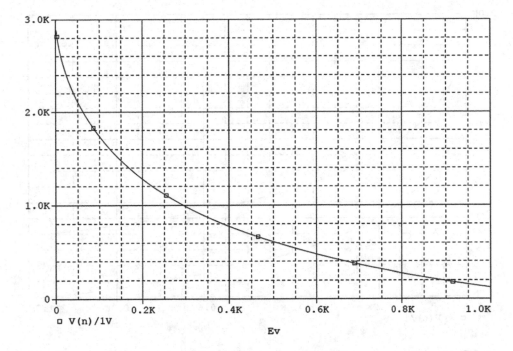

Fig. 3.8 Speed as a function of illuminance at $U_B = 9$ V and $M = 0$

The photodiodes are used as fast photodetectors in the field of measurement, control and regulation.

Data Sheet

BPX90 photodiode, Siemens [6].

Short-circuit current $I_K = 45$ µA at $E_v = 1$ klx, standard light A, $T_f = 2856$ K, (1 klx $\triangleq$ 4.75 mW/cm^2).

Open circuit voltage $U_0 = 450$ mV at $E_v = 1$ klx, standard light A, (1 mW/cm^2 $\triangleq$ 210 lx).

Photosensitivity $S = I_p/E_v = 45$ nA/lx at $U_R = 5$ V, standard light A, $S = I_p/(E_e\ A)$, $A = 5.5$ mm^2.

The saturation current I_S is calculated via I_K and U_0 with Eq. (3.18)

$$I_S = \frac{I_K}{\exp\left(U_0/U_T\right)} \tag{3.18}$$

Here, the temperature stress U_T is according to Eq. (3.19)

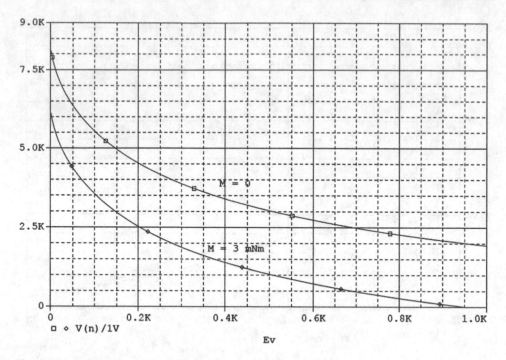

Fig. 3.9 Speed as a function of illuminance at $U_B = 20$ V with M as parameter

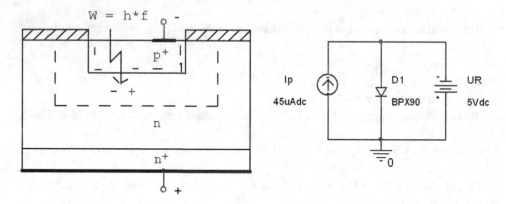

Fig. 3.10 Structure and circuit of the silicon photodiode

$$U_T = \frac{k \cdot T}{e} \tag{3.19}$$

with $k = 1.3806226{-}10^{-23}$ Ws/K and e $= 1.6021918{-}10^{-19}$ As, $U_T = 25.864$ mV at $T = 27\ °C = 300.15$ K. The saturation current is obtained according to Eq. (3.18) with $I_S = 1.25$ **pA** (for $N = 1$).

Parameter Extraction with MODEL EDITOR

Table 3.2 shows value pairs from the blocking characteristic of data sheet [6].

The evaluation with the program MODEL EDITOR yields the reverse saturation current $I_{SR} = 1.52$ nA and the corresponding coefficient $N_R = 5$.

Table 3.3 contains pairs of values of the capacitance characteristic curve from the data sheet.

Using MODEL EDITOR, we obtain the zero-voltage capacitance $C_{JO} = 436$ pF and the exponent $M = 0.433$, as well as the diffusion voltage $V_J = 0.39$ V. Furthermore, the breakdown voltage according to the data sheet has the value $BV = 32$ V.

Task: Characteristic Diagram of the BFX90 Diode

For the photodiode BPX 90 the characteristic $I(U) = f(U)$ with the values of illuminance $E_v = (200, 400, 600, 800, 1000)$ lux shall be plotted as parameters for $U = -5$ V to 0.5 V, see Fig. 3.11.

For $E_v = 1000$ lx, the photocurrent is $I_p = 45$ μA. It is $I_p \sim E_v$. Consequently, $I_p = 9$ μA for $E_v = 200$ lx, $I_p = 18$ μA for $E_v = 400$ lx and continuous.

Analysis

- PSpice, Edit Simulation Profile
- Simulation Settings – Fig. 3.11: Analysis
- Analysis type: DC Sweep
- Options: Primary Sweep
- Sweep variable: Voltage Source
- Name: U
- Sweep type: Linear
- Start value: −5 V
- End value: 0.5 V
- Increment: 1 mV
- Options: Secondary Sweep
- Sweep variable: Current Source

Table 3.2 Data of the blocking characteristic of the photodiode BFX90

U_R in V	2	3.5	5.2	7	10	13	16	18.3	21
I_R in nA	1	2	3	4	5	6	7	8	9

Table 3.3 Data of the capacitance characteristic of the photodiode BFX90

U_R in V	0.01	0.1	0.2	1	2	4	10	30
C_j in pF	430	395	365	255	195	150	110	70

Fig. 3.11 Simulation circuit for
the current-voltage characteristic
of the BPX90 diode

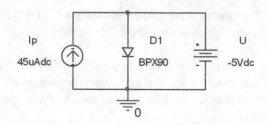

- Name: Ip
- Sweep type: Linear
- Start value: 9 uA
- End value: 45 uA
- Increment: 9 uA
- Apply: OK
- PSpice, run

A Diode **Dbreak** is to be converted to Diode BPX90 via Edit, PSPICE Model with the
previously extracted model parameters as follows:

.model BPX90 D (IS = 1.25p, ISR = 1.52n, NR = 5, CJO = 436p, M = 0.433, VJ = 0.39,
BV = 32).

The analysis result in Fig. 3.12 in the range from V_U = −5 V to 0 V is the characteristic
curve part for diode operation and from V_U = 0 to 0.5 V the part for element operation.

Task: Characteristic Curve Field of a Light Meter
For the light meter according to Fig. 3.13 the characteristic curve field together with the to
represent resistance lines. The voltage U_A is a measure of the illumination.

Analysis

- PSpice, Edit Simulation Profile
- Simulation Settings – Fig. 3.13: Analysis
- Analysis type: DC Sweep
- Options: Primary Sweep
- Sweep variable: Voltage Source
- Name: U
- Sweep type: Linear
- Start value: 0
- End value: 0.6 V
- Increment: 10 uV
- Options: Secondary sweep

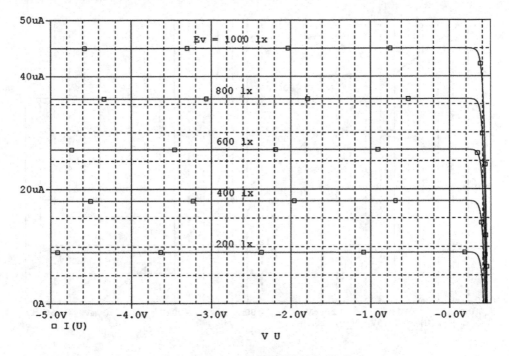

Fig. 3.12 Simulated characteristic diagram of the BPX90 photodiode

Fig. 3.13 Circuit of an
exposure meter

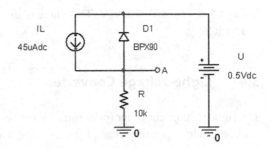

- Sweep variable: Current Source
- Name: IL
- Start value: 9 uA (Ev = 200 lx)
- End value: 45 uA (Ev = 1000 lx)
- Increment: 9 uA
- Apply: OK
- PSpice, run

From the analysis result shown in Fig. 3.14, it can be seen that a voltage drop across the diode of $U_D = 0.14$ V occurs at $E_v = 800$ lx. Thus, the output voltage reaches $U_A = U -$

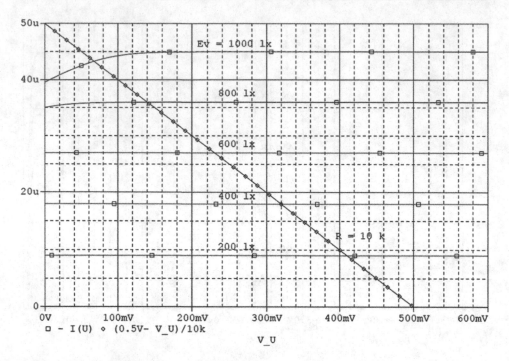

Fig. 3.14 Characteristic diagram of the BPX90 photodiode and resistance curve

$U_D = 0.5\,\text{V} - 0.14\,\text{V} = 0.36\,\text{V}$. This value is also confirmed by the operating point analysis of PSPICE.

3.4   Light-Voltage Converter

The light-voltage converter (principle: *I-V converter*, transimpedance amplifier) consists of a photodiode, operational amplifier and feedback resistor as shown in Fig. 3.15 and converts the photocurrent into an output voltage.

The photocurrent I_p points out of the resistor R, therefore the output voltage in Eq. (3.20) has the positive sign with:

$$U_A = I_p \cdot R \qquad\qquad (3.20)$$

Data Sheet
TSL250R light-to-voltage converter, Texas Instruments, TAOS, AMS [7].
The electrical characteristics of this transducer are shown in Table 3.4.

Fig. 3.15 Schematic diagram of the TSL250R light-voltage converter

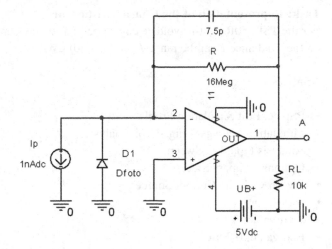

Fig. 3.16 Illustration of the light-voltage converter when using an H-source

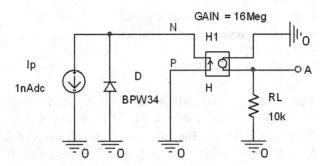

Table 3.4 *I-V converter* TSL250R at $U_{B+} = 5$ V, $T_A = 25$ °C, $\lambda_p = 635$ nm, $R_L = 10$ kΩ

Parameter	Measurement conditions	Min	Type	Max	Unit
U_A output voltage	$E_e = 14.6$ µW/cm²	1.5	2	2.5	V
N_e radiation sensitivity	$\lambda_p = 635$ nm	137			mV/(µW/cm²)
	$\lambda_p = 880$ nm	127			mV/(µW/cm²)

The circuit shown in Fig. 3.15 corresponds to a circuit with a **current-controlled voltage source H** whose parameter GAIN has the dimension of a resistor, see Fig. 3.16. A Si PIN diode BPW34 was used as the photodiode.

If we increase the photocurrent I_p, then the output voltage $U_A = I_p *$GAIN increases linearly. With GAIN $= 16$ MΩ, $U_A = 2$ V at $I_p = 125$ nA. For $U_A = 2$ V, the irradiance is $E_e = 14.6$ µW/cm², see Table 3.4. (1 µW/cm² corresponds to $E_v = 0.21$ lx, 14.6 µW/cm² $= 3.066$ lx).

Thus, it is $E_e = x - I_p$ with $x = E_e/I_p = 14.6$ µW/cm²/125 nA $= 1.168 - 10^8$ µW/(cm² A).

Furthermore, $N_e = U_A/E_e = 2000$ mV/14.6 µW/cm² $= 137$ mV/(µW/cm²), see Table 3.4.

Task: Representation of the Characteristic Curve

For the TSL250R light-to-voltage converter, analyze the dependence of the output voltage on the irradiance E_e in the range $E_e = (0.1-30)$ µW/cm^2.

Analysis

- PSpice, Edit Simulation Profile
- Simulation Settings – Fig. 3.16: Analysis
- Analysis type: DC Sweep
- Options: Primary Sweep
- Sweep variable: Current Source
- Name: Ip
- Start value: 0
- End value: 0.2 uA
- Increment: 1 nA
- Plot, Axis Settings, Axis Variable
- Trace Expression: I_Ip∗1.168E08/1A
- Apply. OK
- PSpice, run

The abscissa and the ordinate are divided logarithmically as in the manufacturer's data sheet. The analysis result of Fig. 3.17 agrees with the data sheet. At $U_A = 2$ V, $E_e = 14.6$ µW/cm^2. The two typical specifications of Table 3.4 with $U_A = 2$ **V** and $N_e = U_A/E_e = $ **137 mV/cm^2** at $\lambda_p = 635$ nm are satisfied.

Task: Switching Times

In the circuit shown in Fig. 3.18, the photodiode is to be pulsed with $I_2 = 125$ nA corresponding to the irradiance $E_e = 14.6$ µW/cm^2. The cathode is connected to the N input of a CMOS operational amplifier, which is constructed with a voltage-controlled voltage source E ($GAIN = 5-10^4$), an input resistance $R_e = 100$ GΩ and an output resistance $R_a = 2$ kΩ. The rise and fall times are to be determined.

Analysis

- PSpice, Edit Simulation Profile
- Simulation Settings – Fig. 3.18: Analysis
- Analysis type: Time Domain (Transient)
- Options: General Settings
- Run to time: 2 ms
- Start saving data after: 1 us
- Maximum step size:10 us
- Apply: OK
- PSpice, run

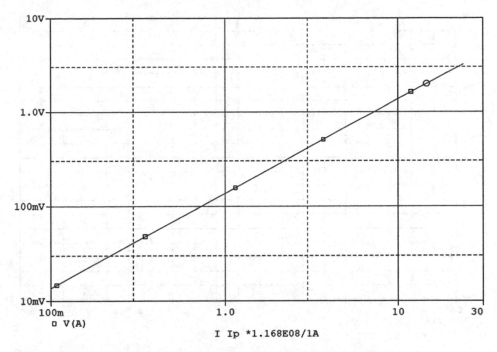

□ V(A)

I Ip *1.168E08/1A

Fig. 3.17 Voltage U_A as a function of irradiance E_e in $\mu W/cm^2$ at $\lambda_p = 635$ nm

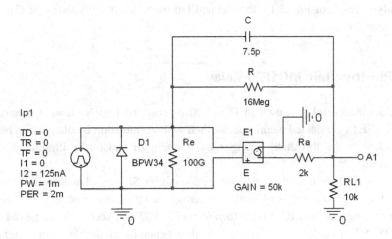

Fig. 3.18 Light-voltage converter using an E-source

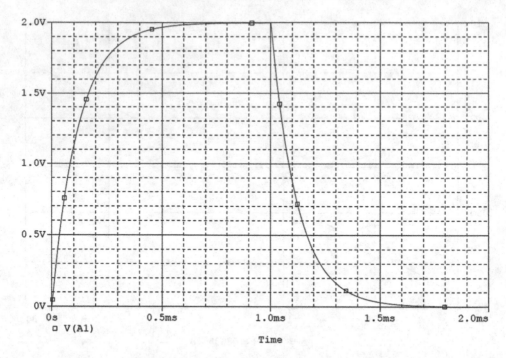

Fig. 3.19 Determination of rise and fall time at $U_B = 5$ V and $\lambda_p = 635$ nm

In the analysis result of Fig. 3.19, the rise and fall time values are satisfied at $C = 7.5$ pF with $t_r = t_f \approx 260$ µs.

3.5   Photovoltaic MOSFET Relay

The light-controlled relay shown in Fig. 3.20 consists of two N-channel enhancement power MOSFETs connected against each other, whose gate-source voltages can be raised to values above the threshold voltages by a chain of pulse-like illuminated silicon photoelements.

A GaAs LED illuminates the series connection of nine Si photoelements (of the type like BPW34). A voltage of about 0.4 V builds up across each photoelement. In the example, the gate-source voltage of the MOSFET is then about 3.6 V. This exceeds the threshold voltage of the MOSFET at the level of $V_{TO} = 2.8$ V, thus generating a usable drain current.

Model the diodes in Fig. 3.20 as follows:

.model LED_red D IS = 1.2E-20 N = 1.46 RS = 2.4 EG = 1.95
. model Dp D IS = 15.46p N = 1 RS = 0.1 ISR = 0.6n CJO = 70p EG = 1.11.

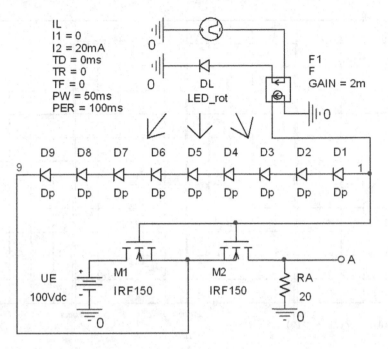

Fig. 3.20 Basic wiring of the photovoltaic relay

You get, for example:

$$U_F = U_T \cdot \ln\left(\frac{I_F}{I_S}\right) = 25,864 \ mV \cdot \ \ln\left(\frac{2 \cdot \ 10^{-3} \ \cdot \ 20 \ mA}{15,46 \ pA}\right) = 0,382 \ \text{V}.$$

With nine diodes, this makes the gate voltage $U_{GS} = 9 - U_F = 3.44$ V.

Task: Pulsed Operation
The photodiodes are to be illuminated with light pulses over a period of 0–200 ms. The output voltage and the voltage across the diodes are to be displayed.

Analysis

- PSpice, Edit Simulation Profile
- Simulation Settings – Fig. 3.20: Analysis
- Analysis type: Time Domain (Transient)
- Options: General Settings
- Run to time: 200 ms
- Start saving data after: 0
- Maximum step size:50 us

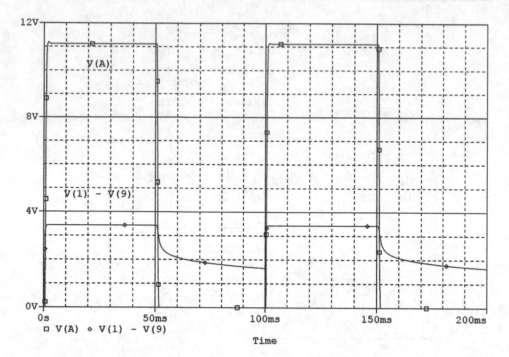

Fig. 3.21 Generated gate-source voltage and output voltage at the photo relay

- Apply: OK
- PSpice, run

The analysis result according to Fig. 3.21 shows the output voltage as well as the voltage across the photoelement chain. With longer selected times, this total voltage decreases to 0 volts during the switching pauses.

The current through the terminating resistor R_A is $I(R_A) = U_A/R_A \approx 11.1$ V/ 20 $\Omega = 0.555$ A.

3.6   RGB Colour Sensor

One possible embodiment of a color sensor contains three silicon diodes accommodated in a common housing, which are provided with a filter in red, green and blue, respectively. The filters are adapted to the color perception of the human eye.

A color sensor of the type S7505-01 from Hamamatsu is considered. The characteristics of this sensor are shown in Table 3.5.

The sensitivity S is obtained either as $S = \frac{I_p}{E_v}$ in nA/lx or in the form $S = \frac{I_p}{E_e \cdot A}$ in A/W. Where E_v is the illuminance in lux and E_e is the irradiance in mW/cm^2.

The saturation current I_S follows from Eq. (3.21) with

Table 3.5 Characteristics of the S7505-1 color sensor from Hamamatsu [8]

Color	Red	Green	Blue
Photosensitive area A	2.25 mm^2	2.25 mm^2	4.50 mm^2
Spectral range λ	(590–720) nm	(480–600) nm	(400–540) nm
Wavelength for maximum sensitivity λ_p	620 nm	540 nm	460 nm
Sensitivity S_{max} at $\lambda = \lambda_p$	0.16 A/W	0.23 A/W	0.18 A/W
Wavelength FWHM for $S = S_{max}/2$	70 nm	60 nm	90 nm
Capacity C_j at $U_R = 0$ V	80 pF	80 pF	150 pF
Short-circuit current I_K at $E_v = 1$ klx	630 nA	490 nA	580 nA

Table 3.6 Values of the dark characteristic curve of the color sensor S7501-01

U_R in V	0.01	0.03	0.1	0.3	1	3	10
I_{RD} in pA	0.85	1.9	3.0	4.0	8.0	19.5	75

Table 3.7 Values of the capacitance characteristic curve of the color sensor S7505-01

U_R in V	0.1	0.3	1	3	10
C_j in pF for D$_{rot}$ and D$_{grün}$	69	59	43	29.5	18
C_j in pF for D$_{blau}$	155	120	86	58	35

$$I_S = \frac{I_K}{e^{U_0/U_T}} \qquad (3.21)$$

With an assumed value of the open circuit voltage $U_0 = 350$ mV is obtained:

$I_S = 0.84$ pA for D$_{rot}$, $I_S = 0.77$ pA for D$_{grün}$ and $I_S = 0.65$ pA for D$_{blau}$.

The reverse saturation current I_{SR} and the associated emission coefficient N_R can be determined from the manufacturer's dark characteristic $I_{RD} = f(U_R)$. Table 3.6 shows the pairs of values taken from the characteristic curve.

Using MODEL EDITOR, extract the model parameters $I_{SR} = 19.41$ pA and $N_R = 5$.

Table 3.7 shows the dependence of the junction capacitance C_j on the reverse voltage U_R.

With the program MODEL EDITOR you extract the following model parameters:

$C_{JO} = 76.06$ pF, $M = 0.442$ and $V_J = 0.3905$ for diodes D$_{rot}$ and D$_{grün}$ and
$C_{JO} = 168$ pF, $M = 0.506$ and $V_J = 0.3905$ for diode D$_{blau}$.

From this follows the SPICE models of the three diodes with

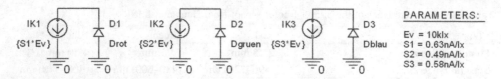

Fig. 3.22 Circuit for displaying the short-circuit currents of the color sensor 7506-1

. model Drot D IS = 0.84p ISR = 19.41p NR = 5 CJO = 76.06p M = 0.442 VJ = 0.3905
. model Dgruen D IS = 0.77p ISR = 19.41p NR = 5 CJO = 76.06p M = 0.442
   VJ = 0.3905
. model Dblue D IS = 0.65p ISR = 19.41p NR = 5 CJO = 168p M = 0.506 VJ = 0.3905

The circuit shown in Fig. 3.22 is used to display the short-circuit currents of the three diodes. The sensitivity is $S = I_K/E_v$ is a parameter, see also Table 3.5. If the cathodes are connected to ground, the direction of the photocurrents must be changed.

Task: Dependence of the Short-Circuit Currents on the Illumination Intensity
The dependence of the short-circuit currents of the three diodes as a function of the illuminance in the range $E_v = 10$ lx to 10 klx is to be analyzed.

Analysis

- PSpice, Edit Simulation Profile
- Simulation Settings – Fig. 3.22: Analysis
- Analysis type: DC Sweep
- Options: Primary sweep
- Sweep variable: Global Parameter
- Parameter Name: Ev
- Sweep type: Logarithmic Decade
- Start value: 10
- End value: 10 k
- Points/Decade: 100
- Apply: OK
- PSpice, run

The analysis result according to Fig. 3.23 corresponds to the data sheet specification.
   The dark characteristic can be simulated with the circuit shown in Fig. 3.24.

Task: Dark Characteristic
To be analyzed is the dark characteristic $I_R = f(U_R)$ at $E_v = 0$ for $U_R = 0.01$ V to 10 V on a logarithmic scale.

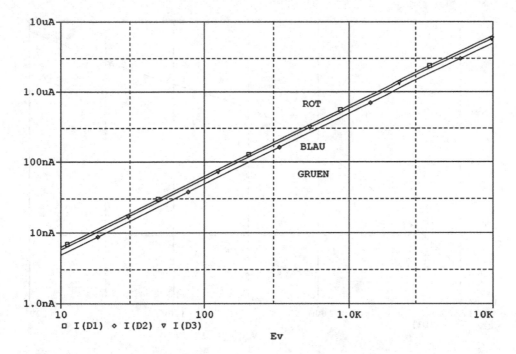

Fig. 3.23 Short-circuit currents as a function of the illuminance of the color sensor

Fig. 3.24 Circuit for simulating
the dark characteristic curve

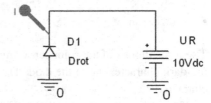

Analysis

- PSpice, Edit Simulation Profile
- Simulation Settings – Fig. 3.24: Analysis
- Analysis type: DC Sweep
- Options: Primary Sweep
- Sweep variable: Voltage Source
- Name: UR
- Sweep type: Logarithmic Decade
- Start value: 10 mV
- End value: 10 V
- Points/Decade. 100

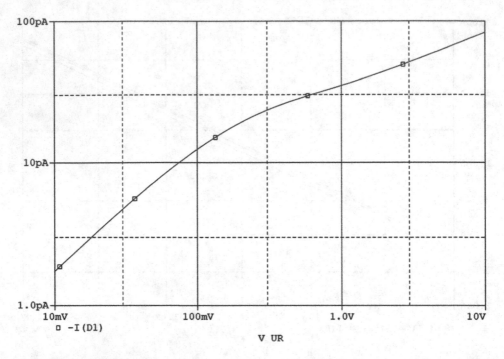

100pA

10pA

1.0pA

10mV   100mV   1.0V   10V

□ -I(D1)

V_UR

Fig. 3.25 Simulated dark characteristic of the color sensor S7505-01

- Apply: OK
- PSpice, run

The three diodes of the color sensor have identical values for I_{SR} and N_R. Therefore, only the dark characteristic of the diode D_{rot} is simulated. The manufacturer only specifies one characteristic curve.

The analysis result according to Fig. 3.25 is within the range of the values of the dark characteristic curve of the manufacturer, but deviations occur in the middle range.

Task: Capacity Characteristics

The capacitance characteristics $C_j = f(U_R)$ of the three diodes of the colour sensor for $U_R = 0.1$ V to 10 V.

Equation (3.22) describes the relationship between the reverse voltage U_R and the junction capacitance C_j.

$$U_R = V_j \cdot \left[\left(\frac{C_{JO}}{C_j} \right)^{\frac{1}{M}} - 1 \right] \tag{3.22}$$

Fig. 3.26 Circuit for simulating
the capacitance characteristic for
diode D_{rot}

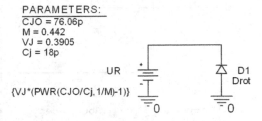

PARAMETERS:
CJO = 76.06p
M = 0.442
VJ = 0.3905
Cj = 18p

UR

{VJ*(PWR(CJO/Cj, 1/M)-1)}

D1
Drot

The circuit for simulating the capacitance characteristic $C_j = f(U_R)$ for diodes D_{rot} and D_{gruen} is shown in Fig. 3.26. Both diodes have identical model parameters for the initial capacitance C_{JO}, the exponent M and the diffusion voltage V_J valid at the reverse voltage $U_R = 0$ V.

Equation (3.22) is obtained by transforming the relation: $C_j = C_{JO}/(1 + U_R/V_J)^M$.

Analysis

- PSpice, Edit Simulation Profile
- Simulation Settings – Fig. 3.26: Analysis
- Analysis type: DC Sweep
- Options: Primary sweep
- Sweep variable: Global Parameter
- Parameter Name: Cj
- Sweep type: Linear
- Start value. 18p
- End value. 80p
- Increment: 0.01p
- Apply: OK
- PSpice, run
- Edit diagram:
- Trace, Add trace: Cj, OK
- Plot, Axis Settings, Axis variable
- Simulation output variable: V(UR:+)
- Plot Axis settings
- X-Axis, User defined: 0.1 V to 10 V, Log, OK
- Plot Axis Settings
- Y-Axis, User defined: 10p to 200p, Log, OK

The analysis result shown in Fig. 3.27 shows the decrease in junction capacitance with increasing reverse voltage.

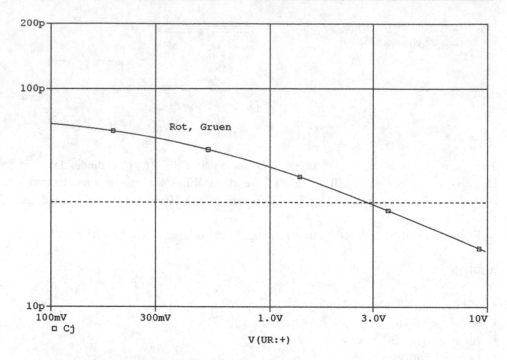

Fig. 3.27 Identical capacity characteristics for D_{rot} and D_{gruen}

Fig. 3.28 Circuit for simulating
the capacitance characteristic of
diode D_{blau}

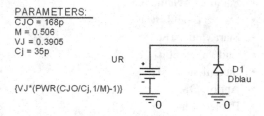

PARAMETERS:
CJO = 168p
M = 0.506
VJ = 0.3905
Cj = 35p

{VJ*(PWR(CJO/Cj, 1/M)-1)}

Figure 3.28 shows the circuit for simulating the capacitance characteristic of diode
D_{blau}. This diode has twice as large a photosensitive area as the other two diodes, see
Table 3.5. Its capacitance is therefore higher.

The analysis of the capacitance characteristic of the diode D_{blau} is carried out in the same
way as for the diode D_{rot}, but with.

Start value: 35p, End value: 168p and Increment: 0.01p.

Figure 3.29 shows the dependence of the junction capacitance on the reverse voltage for
diode D_{blau}.

The simulated capacitance characteristics of the three diodes of the color sensor S7505-
01 according to Figs. 3.27 and 3.29 correspond to those of the data sheet.

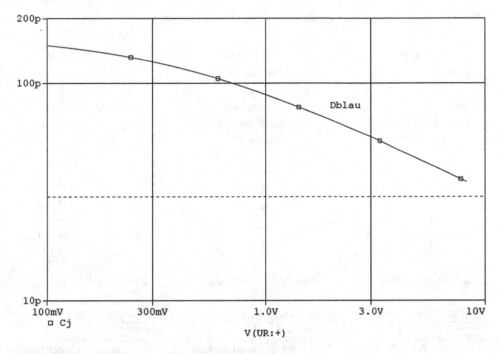

Fig. 3.29 Capacitance characteristic of the blue diode

Task: Sensitivity as a Function of Wavelength

The dependence of the sensitivity S in A/W on the wavelength λ in nm can be approximated by a modified bell curve according to Eq. (3.23).

$$S = \frac{S_{\max}}{\exp\left[\frac{\ln(2) \cdot (\lambda - \lambda_p)^2}{(FWHM/2)^2}\right]} \tag{3.23}$$

For simplicity, coefficients are introduced for Eq. (3.23). The coefficient k_1 is obtained from Eq. (3.24) to give

$$k_1 = (\lambda - \lambda_p) \tag{3.24}$$

and the coefficient k_2 follows from Eq. (3.25) with

$$k_2 = \frac{\ln(2)}{(FWHM/2)^2} \tag{3.25}$$

Table 3.8 Characteristic values for the dependence of the sensitivity on the wavelength

D_{blau}	D_{gruen}	D_{rot}
$S_{maxb} = 0.18$ A/W	$S_{maxg} = 0.23$ A/W	$S_{maxr} = 0.16$ A/W
$\lambda_{pb} = 460$ nm	$\lambda_{pg} = 540$ nm	$\lambda_{pr} = 620$ nm
$b_1 = (\lambda - \lambda_{pb})$	$g_1 = (\lambda - \lambda_{pg})$	$r_1 = (\lambda - \lambda_{pr})$
$FWHM_b = 90$ nm	$FWHM_g = 60$ nm	$FWHM_r = 70$ nm
$b_2 = \ln(2)/(FWHM_b/2)^2$	$g_2 = \ln(2)/(FWHM_g/2)^2$	$r_2 = \ln(2)/(FWHM_r/2)^2$
$b_2 = 3.4229{-}10^{14}$ m^{-2}	$g_2 = 7.7016{-}10^{14}$ m^{-2}	$r_2 = 5.6583{-}10^{14}$ m^{-2}

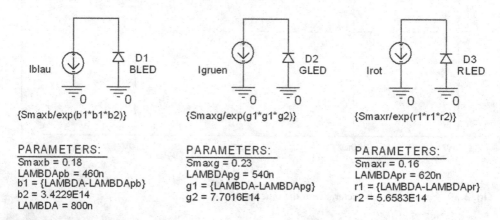

Fig. 3.30 Circuits for the dependence of sensitivity on wavelength

Table 3.8 shows the characteristic values for the three diodes of the color sensor, which are derived from Table 3.5. Instead of the general coefficients k_1 and k_2, the respective diode parameters for blue, green and red apply: b_1 and b_2 or g_1 and g_2 as well as r_1 and r_2.

The expression for the spectral sensitivity S in the unit A/V according to Eq. (3.23) is entered as the value of the luminous flux, enclosed in curly brackets, for the respective diode. The circuits for simulating the function $S = f(\lambda)$ are shown in Fig. 3.30.

Analysis

- PSpice Simulation Profile
- Simulation Settings – Fig. 3.30: Analysis
- Analysis type: DC Sweep
- Options: Primary sweep
- Sweep variable: Global Parameter
- Parameter Name: LAMBDA
- Sweep type: Linear

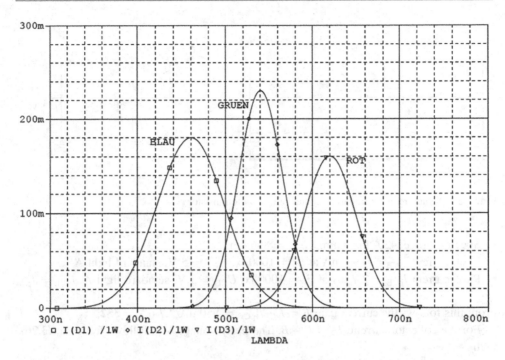

Fig. 3.31 Sensitivity in mA/W as a function of wavelength LAMBDA in nanometers

- Start value: 300 n
- End value: 800 n
- Increment: 0.1 n
- Apply, OK
- PSpice, run

The analysis result of Fig. 3.31, which is based on Eq. (3.23), shows good agreement with the data sheet specification for the S7505-01 color sensor manufactured by Hamamatsu.

3.7 Phototransistor

Phototransistors have on the one hand a higher photosensitivity but on the other hand higher switching times than photodiodes. They are manufactured as npn planar transistors and are suitable for use in threshold switches, optocouplers and light barriers. Figure 3.32 shows the basic structure of the phototransistor together with a circuit for simulating characteristic curves with type BP103-4.

Data Sheet
BP 103 phototransistor, group 4 of photosensitivity [9].

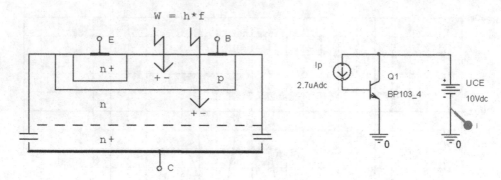

Fig. 3.32 Structure and simulation circuit of the phototransistor

- Spectral range: 420–1130 nm, $\lambda_{Smax} = 850$ nm
- Photocurrent $I_{pCE} = 0.95$ mA at $E_v = 1000$ lx, $U_{CE} = 5$ V, standard light A
- Photocurrent $I_{pCB} = 2.7$ μA at $E_v = 1000$ lx, $U_{CB} = 5$ V, standard light A, $I_{pCB} = I_p$

From this follows the current gain $B = I_{pCE}/I_{pCB} = 950$ μA/2.7 μA = **352**.

For the collector current, $I_C \approx I_p - B$. The Early voltage V_{AF} follows from Eq. (3.26) with

$$V_{AF} = \frac{U_{CE2} \cdot I_{C1} - U_{CE1} \cdot I_{C2}}{I_{CE2} - I_{CE1}} \tag{3.26}$$

From the characteristic curve $I_C = f(U_{CE})| \; I_B$ of the data sheet, we take the value pairs at $I_B = 1.6$ μA: $U_{CE1} = 10$ V, $I_{C1} = 0.84$ mA $U_{CE2} = 50$ V, $I_{C2} = 0.93$ mA.

Using Eq. (3.26), V is then $_{AF} \approx$ **360 V**. A transistor QbreakN is remodeled via Edit, PSPICE Model with

. model BP103_4 NPN BF = 352 VAF = 360 IS = 10f.

Task: Output Characteristic Field
The characteristic field $I_C = f(U_{CE})$ with E_v as parameter for U_{CE} from 0 to 10 V and $E_v = (200, 400, 600, 800, 1000)$ lx is to be represented. $E_v \sim I_p$ applies. The circuit shown in Fig. 3.32 is to be used.

Analysis

- PSPICE, Edit Simulation Profile
- Simulation Settings – Fig. 3.32
- Analysis type: DC Sweep
- Options: Primary sweep

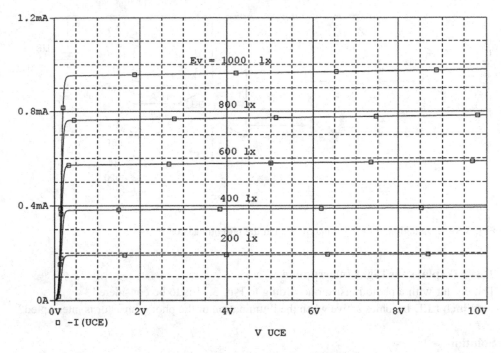

Fig. 3.33 Simulated characteristic diagram of phototransistor BP103-4

- Sweep variable: Voltage Source
- Parameter Name: UCE
- Sweep type: Linear
- Start value: 0
- End value: 10 V
- Increment: 1 mV
- Options: Secondary Sweep
- Sweep variable: Current Source
- Name: Ip
- Sweep type: Linear
- Start value: 0.54 uA
- End value. 2.7 uA
- Increment: 0.54 uA
- Apply: OK
- PSpice, run

The analysis leads to Fig. 3.33 and shows that the transistor BP103-4 approximately achieves the collector current $I_{PCE} = 0.95$ mA at $E_v = 1000$ lx and $U_{CE} = 5$ V as specified in the data sheet table.

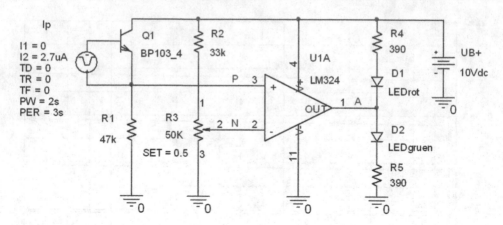

Fig. 3.34 Evaluation of the light interruption at the phototransistor

Task: Displays for Light Interruption

The circuit with a phototransistor according to Fig. 3.34 is to be considered.

Which LED becomes active when the illumination of the phototransistor is interrupted?

Solution

If the illumination of the phototransistor is interrupted, then $U_P < U_N$. The output voltage U_A gets to LOW and LED_{rot} becomes active while LED_{gruen} is off [2, 10].

Analysis

- PSpice, Edit Simulation Profile
- Simulation Settings – Fig. 3.34: Analysis
- Analysis type: Time Domain (Transient)
- Options: General Settings
- Run to time: 8 s
- Start saving data after: 0
- Maximum step size: 10 ms
- Apply: OK
- PSpice, run

With the analysis result shown in Fig. 3.35, the predictions about the voltages at the comparator and the LED currents are confirmed.

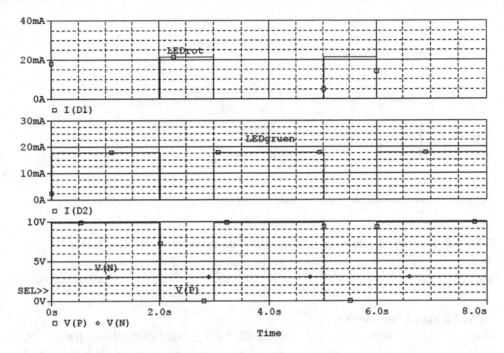

Fig. 3.35 Time dependence of voltages and LED currents

3.8   Fork Coupler

Figure 3.36 shows the structure and simulation circuit for the fork coupler. The GaAs transmitter diode transmits its light across the fork gap to the base collector diode of the silicon npn phototransistor. The light can be interrupted in the fork.

Data Sheet
Fork coupler EE_SX1106 from OMRON [11].

- Limits: $I_{Fmax} = 50$ mA, $I_{Cmax} = 30$ mA, $U_{CE0} = 30$ V, $P_C = 80$ mW.
- Typical characteristic value: $U_F = 1.3$ V at $I_F = 50$ mA.

Table 3.9 shows the pairs of values for the forward characteristic of the GaAs diode taken from the data sheet.

From the input of the above data into the MODEL EDITOR program, the model parameters of the transmitting diode follow with $I_S = \mathbf{76.44}$ **p**, $N = \mathbf{2.349}$, $R_S = \mathbf{1.438}$ $\boldsymbol{\Omega}$.

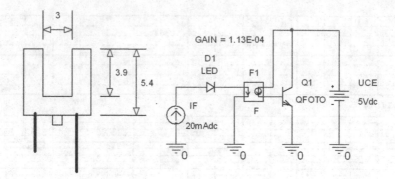

Fig. 3.36 Dimensions and wiring of the fork coupler

Table 3.9 Data from the characteristic $I_F = f(U_F)$ of the transmitter diode at 25 °C [11]

U_F/V	1.07	1.10	1.14	1.21	1.25	1.28	1.30
I_F/mA	3	5	10	20	30	40	50

Task: Passing Characteristic

Using the circuit shown in Fig. 3.37, analyze the LED forward characteristic for −25 °C, 25 °C, and 75 °C. For this purpose, a diode Dbreak is to be modelled via Edit, PSPICE Model as follows:

.model LED D (IS = 76.44p, N = 2.349, RS = 1.438, EG = 1.6)

The parameter EG (Energy Gap) determines the temperature dependence in addition to IS.

Analysis

- PSpice, Edit Simulation Profile
- Simulation Settings – Fig. 3.37: Analysis
- Analysis type: DC Sweep
- Options: Primary Sweep
- Sweep variable: Voltage Source
- Name: UF

Fig. 3.37 Circuit for simulating
the forward characteristic curve

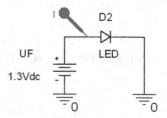

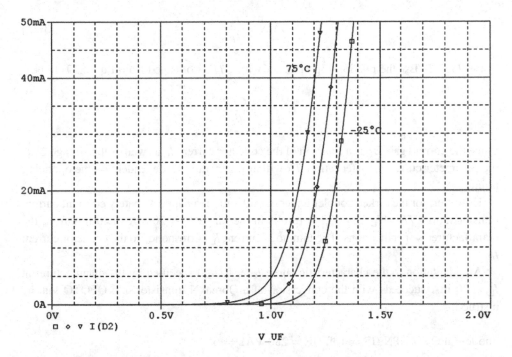

Fig. 3.38 Simulated forward characteristic of the emitting diode at −25 °C, 25 °C and 75 °C

- Sweep type: Linear
- Start value: 0
- End value: 1.8 V
- Increment: 1 mV
- Options: Temperature (Sweep)
- Repeat the simulation for each of the temperatures: −25, 25, 75
- Apply: OK
- PSpice, run

The analysis result according to Fig. 3.38 shows an approximate agreement of the simulated forward characteristics with the data sheet specification for the specified temperatures. An important parameter of the fork coupler is the current transfer ratio (*CTR*) according to Eq. (3.27).

$$CTR = \frac{I_C}{I_F} \tag{3.27}$$

For the parameter *GAIN* of the current controlled current source F applies according to Eq. (3.28):

$$GAIN = \frac{I_B}{I_F} \qquad (3.28)$$

Using $I_B = I_C/B_N$, the relationship of *GAIN* with *CTR* is obtained from Eq. (3.29) to be

$$GAIN = \frac{CTR}{B_N} \qquad (3.29)$$

Since the current gain B_N is a function of the collector current I_C as well as the voltage U_{CE}, the parameter current transfer ratio *CTR* will also generally be dependent on the operating point.

However, for the forked coupler under consideration, an approximately constant current transfer ratio $CTR = I_L /I_F \approx 3.5\%$ results in the range $I_F \approx 20$ mA to 40 mA from the manufacturer's specifications. The luminous current I_L corresponds to the collector current I_C.

An adaptation to the characteristic curve $I_C = f(U_{CE})$ specified by the manufacturer at $I_F = 20$ mA succeeds with the conversion of a QbreakN transistor into QFOTO via the instruction

.model QFOTO NPN (IS = 10f, BF = 280 VAF = 65)

In this case, the slope of the characteristic curve specified in the data sheet $I_L = f(U_{CE})$ for $I_F = 20$ mA an initial value for determining the EARLY voltage V_{AF}. This characteristic was approximated by alternately tuning the values of B_F, V_{AF} and *GAIN*. For a default value $U_{CE} = 7$ V the following follows $B_N = B_F (1 + U_{CE}/V_{AF}) = 310$. It is $B_N = 310 > B_F = 280$, because in the transistor model no track resistors and buckling currents could be inserted, but the EARLY- voltage *VAF* causes an increase of B_N compared to B_F. The value of *GAIN* in Fig. 3.36 is obtained from $GAIN = CTR/B_N = 3.5\%/310 = 1.13{-}10^{-4}$. Because of the large fork width, this value is smaller than that of a common diode-transistor optocoupler.

Task: Output Characteristic Field
Using the circuit shown in Fig. 3.36, simulate the output characteristic $I_C = f(U_{CE})$ of the forked coupler for $U_{CE} = 0$ to 10 V with the forward current of the transmitter diode $I_F = (10, 20, 30, 40, 50)$ mA as a parameter.

Analysis

- PSpice, Edit Simulation Profile
- Simulation Settings – Fig. 3.36: Analysis
- Analysis type: DC Sweep
- Options: Primary Sweep
- Sweep variable. Voltage Source

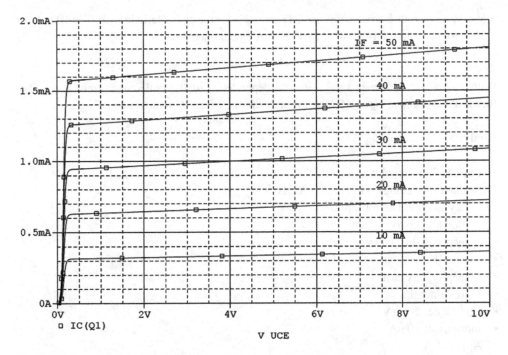

Fig. 3.39 Simulated output characteristic field of the fork coupler EE_SX1106

- Name: UCE
- Sweep type: Linear
- Start value: 0
- End value: 10 V
- Increment: 1 mV
- Options: Secondary Sweep
- Sweep variable: Current source
- Name: IF
- Sweep type: Linear
- Start value: 0
- End value: 50 mA
- Increment: 10 mA
- Apply: OK
- PSpice, run

In the analysis result of Fig. 3.39, the characteristic curve specified by the manufacturer for $I_F = 20$ mA is well reproduced in the range $U_{CE} = 1$ V to 10 V. For $U_{CE} < 1$ V, the phototransistor should be modeled for better agreement with other parameters such as ISE, NE, ISC, NC, RB RC, RE. The output characteristics for $I_F = 10$ mA and $I_F = 50$ mA show increased values compared to the data sheet because they were simulated based on $CTR =$

3.5%. However, their true values are CTR $\approx$ 3.0% for the characteristic with $I_F = 10$ mA or 3.36% for the characteristic with $I_F = 50$ mA as parameter.

Task

With the circuit shown in Fig. 3.36, the dependence of the collector current I_C on the forward current I_F at $U_{CE} = 5$ V at the temperature of 25 °C to simulate.

Analysis

- PSpice, Edit Simulation Profile
- Simulation Settings – Fig. 3.36: Analysis
- Analysis type: DC Sweep
- Options: Primary Sweep
- Sweep variable: Current Source
- Name: IF
- Sweep type: Linear
- Start value: 0
- End value: 50 mA
- Increment: 10 uA
- Apply: OK
- PSpice, run

In the analysis result of Fig. 3.40, the manufacturer's characteristic curve is shown in the range $I_F = 20$ mA to 40 mA well reproduced.

Task: Light Interruption

For the operating point $I_F = 20$ mA, $U_{CE} = 5$ V and $Temp = 25$ °C, the collector current is to be simulated as a function of the distance d, see Fig. 3.41. The manufacturer's characteristic curves show that the light beam is interrupted for horizontal crossing of the fork by means of a separating piece at $d \approx 2$ mm. On the other hand, if the separator dips vertically into the fork from above, then the light beam is blocked at $d \approx 1$ mm. In the parameters, the distances d and d_0 are specified in the unit of millimeters.

The light interruptions can be approximated by the empirical Eq. (3.30) for the current source I_1:

$$I_1 = \frac{I_N}{\exp{(d - d_0)}^{30}} \tag{3.30}$$

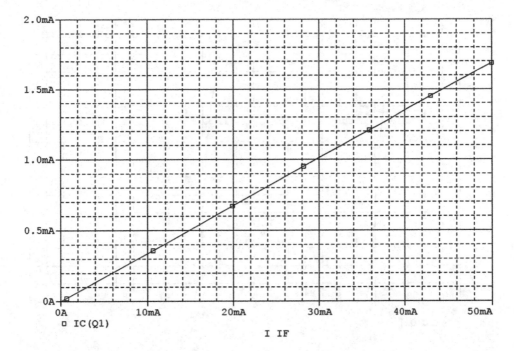

Fig. 3.40 Collector current as a function of forward current of the GaAs diode

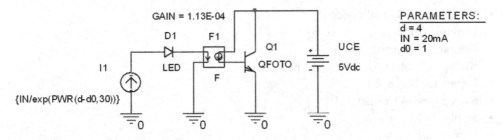

Fig. 3.41 Circuit for displaying a horizontal or vertical light interruption

Analysis

- PSpice, Edit Simulation Profile
- Simulation Settings – Fig. 3.41: Analysis
- Analysis type: DC Sweep
- Options: Primary Sweep
- Sweep variable: Global Parameter
- Parameter Name: d (in millimeters)
- Sweep type: Linear
- Start value: 0.5 (in millimeters)

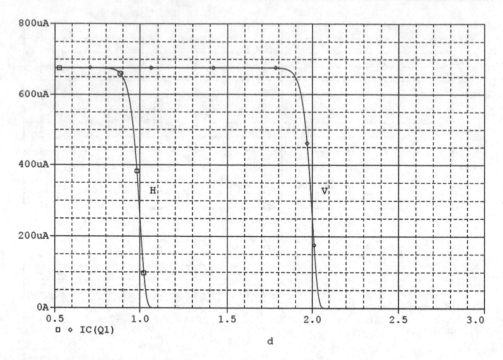

Fig. 3.42 Horizontal or vertical immersion in the fork, distance d in millimetres

- End value: 3 (in millimeters)
- Increment: 1 m (1 m = 1–10-3)
- Options: Parametric sweep
- Sweep variable: Global Parameter
- Parameter Name: d0
- Sweep type: Value list: 0.1
- Apply: OK
- PSpice, run

The analysis result in Fig. 3.42 approximates the manufacturer's specifications for horizontal and vertical light interruption.

Task: Speed Measurement
The speed measurement is to be simulated with the fork coupler, see Fig. 3.43. The interruption of the light beam is simulated with the pulse current source I_P.

Analysis

- PSpice, Edit Simulation Profile
- Simulation Settings – Fig. 3.43: Analysis

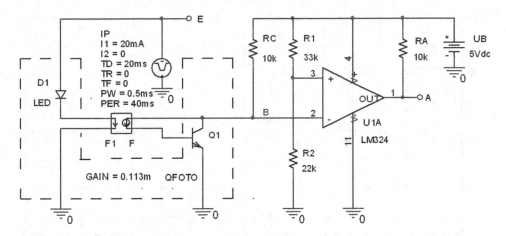

Fig. 3.43 Using the fork coupler for speed measurement

- Analysis type: Time Domain (Transient)
- Options: General Settings
- Run to time: 80 ms
- Start saving data after: 0
- Maximum step size: 10 us
- Apply: OK
- PSpice, run

The analysis result according to Fig. 3.44 shows that the interruption is detected with the output voltage. When the light beam is interrupted with $I_F = 0$, the base current $I_B = 0$, the voltage between collector and emitter reaches $U_{CE} = 5$ V. With $U_P < U_N$ the output voltage $U_A = 0$.

3.9 Reflex Light Barrier

The reflex light barrier contains an IR-GaAs transmitter diode, a Si-npn phototransistor and a daylight cut filter. The circuit according to Fig. 3.45 shows the interaction of these components. The phototransistor is driven by the reflected light from the infrared emitting diode. Depending on the distance x to the reflector, only a small proportion of the emitted light reaches the base collector diode of the receiver. Retro-reflective sensors are used as motion detectors as well as for position detection and speed monitoring.

Data Sheet
SFH 9201 retro-reflective sensor from OSRAM [12].

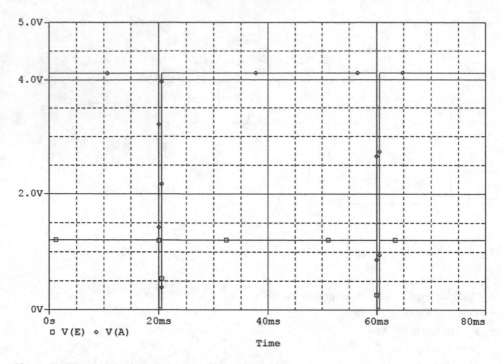

Fig. 3.44 Illustration of the light interruption at the fork coupler

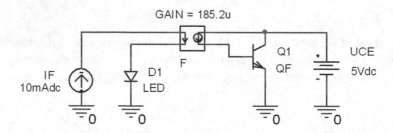

Fig. 3.45 Model of the reflex light barrier for the simulation of characteristic curves

Limit values: $U_R = 5$ V, $I_F = 50$ mA, $U_{CE} = 16$ V, $I_C = 10$ mA.

Characteristic values: $U_F = 1.25$ V at $I_F = 50$ mA, I_{CE} typ $= 0.7$ mA at $I_F = 10$ mA, $U_{CE} = 5$ V, reflector distance: $x = 1$ mm.

Reflector with 90% reflection: Kodak paper, neutral white.

Task: Model Parameters of the Transmitter Diode

At $T_A = 25$ °C, the pairs of values of the forward characteristic follow from the data sheet according to Table 3.10.

Table 3.10 Data sheet information on the transmitter diode of the SFH 9201 reflex light barrier

V_F in V	1.12	1.145	1.18	1.25
I_F in mA	5	10	20	50

Table 3.11 Measured values for the forward characteristic of the transmitter diode according to [13]

V_F in V	0.9482	1.004	1.058	1.102	1.122	1.151	1.184
I_F in mA	0.1	0.3	1	3	5	10	20
V_F in V	1.196	1.207	1.216	1.224	1.231	1.238	–
I_F in mA	25	30	35	40	45	50	–

The data of Table 3.10 are extended by measurements according to Table 3.11.

Parameter extraction with the Model Editor program provides the values:

Saturation current $I_S = \mathbf{18.727}$ **fA**, coefficient $N = \mathbf{1.6466}$, series resistance $\boldsymbol{R_S = 0.3582}$.

Furthermore, from the data sheet follows the value of the reverse current with $I_R = 0.01~\mu A$ at $U_R = 5$ V.

Equation (3.31) applies approximately for the reverse current

$$I_R = I_S + I_{SR} \cdot \left(1 + \frac{U_R}{V_J}\right)^M \tag{3.31}$$

With $U_R = 5$ V, $I_S < <I_{SR}$ and the typical values $V_J = 0.7$ V and $M = 0.33$, the reverse saturation current $\boldsymbol{I_{SR} = 5}$ **nA** is obtained. According to the data sheet, the capacitance is $\boldsymbol{C_{JO} = 25}$ **pF** at $U_R = 0$ V and $f = 1$ MHz. The model parameters also include the value of the energy band gap $\boldsymbol{E_G = 1.5}$ **eV**. With the value of the reverse current emission coefficient $\boldsymbol{N_R = 3.4}$, an acceptable fit to the LED forward characteristics $U_F = f(TEMP)$ with I_F as parameter is subsequently achieved. This set of characteristics is analyzed using the circuit shown in Fig. 3.45. Thus, the transmit LED can be modeled as follows:

.model LED D (IS = 18.727f, N = 1.6466, RS = 0.3582, EG = 1.5, ISR = 5n, VJ = 0.7,
   M = 0.33, + NR = 3.8, CJO = 25p).

Task: Temperature Response of the Forward Voltage

The circuit shown in Fig. 3.45 can be used to analyse the dependence of the LED forward voltage U_F in the range from -40 to $90\,°C$. The forward current can be varied with the LED forward voltage. The forward current is given by $I_F = 5$ mA, 10 mA and 15 mA.

Analysis

- PSpice, Edit Simulation Profile
- Simulation Settings – Fig. 3.45: Analysis
- Analysis type: DC Sweep
- Options: Primary Sweep
- Sweep variable: Temperature
- Sweep type: Linear
- Start value: −40
- End value: 90
- Increment: 1
- Options: Secondary Sweep
- Sweep variable: Current Source
- Name: IF
- Sweep type: Value list: 5 m 10 m 20 m
- Apply: OK
- PSpice, run

With the analysis result according to Fig. 3.46, the LED characteristics of the data sheet are approximately achieved.

Task: Model Parameters of the Phototransistor

In the data sheet, the output characteristic $I_C = f(U_{CE})$ is given with I_F as parameter. This representation is valid for a reflector distance of $x = 1$ mm at 90% reflection. Because of the high transmission losses in reflective photoelectric sensors, the phototransistor receiver is designed for a high ideal current gain $B_F > 400$. The Early voltage V_{AF}, which is actually used to model the slope of the characteristics $I_C = f(U_{CE})$ with the base current I_B as a parameter, also influences the slope of the characteristics with $V_{AF} \approx 100$ V when the LED forward current I_F occurs as a parameter.

With the value $GAIN = 185.2$ u of the current-controlled current source F in the circuit shown in Fig. 3.45, a collector current $I_C = 0.74$ mA is obtained for the distance $x = 1$ mm at 90% reflection and $U_{CE} = 5$ V as well as $I_F = 10$ mA. The data sheet gives a typical value of $I_C = 0.70$ mA, and the characteristic $I_C = f(I_F)$ gives $I_C = 0.75$ mA for these conditions.

From the comparison of the model parameters of other phototransistors in optocouplers and light barriers and via adjustments to the given characteristic curve field of the data sheet of the light barrier SFH 9201, the following parameters for the phototransistor resulted:

.model QF NPN (IS = 2f, BF = 420, VAF = 130, IKF = 0.09, ISE = 25p, NE = 2.9).

Task: Output Characteristic Field

The output characteristics of the phototransistor for $U_{CE} = 0.1$ to 10 V with the parameter $I_F = 5$ mA, 10 mA, 15 mA, 20 mA and 25 mA shall be analysed.

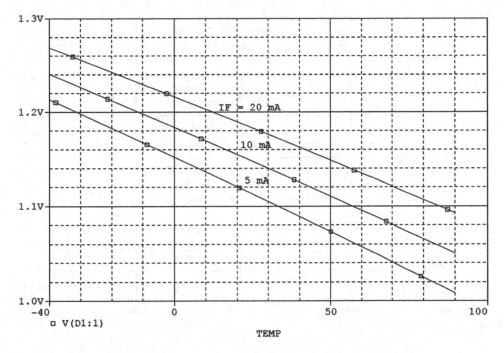

Fig. 3.46 Temperature dependence of forward voltage with current as parameter

Analysis

- PSpice, Edit Simulation Profile
- Simulation Settings – Fig. 3.45: Analysis
- Analysis type: DC Sweep
- Options: Primary Sweep
- Sweep variable: Voltage Source
- Sweep type: Logarithmic, Decade
- Start value: 0.1
- End value: 10
- Points/Decade: 100
- Options: Secondary Sweep
- Sweep variable. Current Source
- Name: IF
- Sweep type: Linear
- Start value: 5 m
- End value: 25 m
- Increment: 5 m
- Apply: OK
- PSpice, run

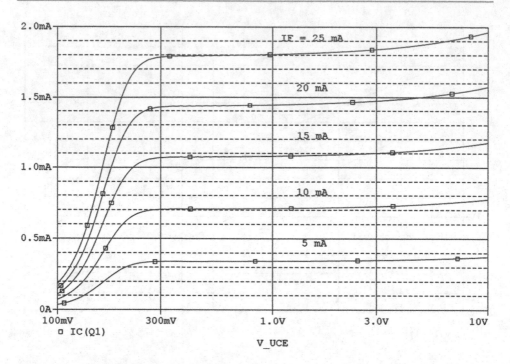

Fig. 3.47 Output characteristic field with the LED current as parameter and $x = 1$ mm

The analysis result according to Fig. 3.47 corresponds approximately to the characteristic curve field of the data sheet.

Task: Collector Current as a Function of the Reflector Distance x
The data sheet also contains a characteristic curve $I_C/I_{Cmax} = f(x)$ using Kodak test paper neutral white for 90% reflection. Table 3.12 shows some pairs of values of this characteristic curve. From $I_C = 0.74$ mA at $I_C/I_{Cmax} = 92\%$ follows $I_{Cmax} = 0.8043$ mA at $I_C/I_{Cmax} = 100\%$. This gives the collector currents for the individual reflector spacings x. The highest flowing collector current $I_{C98\%} = \mathbf{0.788}$ **mA** appears for $x = \mathbf{0.75}$ **mm**. With the given models of the LED and phototransistor, this value is obtained via the operating point analysis for **GAIN$_{98\%}$ = 196.7 u**. This maximum value is valid for $x = 0.75$ mm. The decrease of GAIN for $x < 0.75$ mm or $x > 0.75$ mm is described by Eq. (3.32).

$$k = \frac{GAIN(x)}{GAIN(98\%)} = \frac{a \cdot x}{x^n + b} \tag{3.32}$$

The maximum value $k = 1$ is reached when $a - x = 1$ and $x^n + b = 1$. Given $x = 0.65$ (in mm), $a = 1.5385$ is obtained and given $n = 2.55$, $b = 0.6666$. Since no source FPOLY is available, the current I_{F2} is converted to an equivalent voltage across the resistor R_1 given

Table 3.12 Collector currents as a function of reflector distance x

x in mm	0	0.25	0.50	0.75	1	1.5
I_C / I_{Cmax} in %	0	50	80	98	92	65
I_C in mA	0	0.402	0.643	0.788	0.740	0.523
x in mm	2	2.5	3	3.5	4	4.5
I_C / I_{Cmax} in %	48	34	24,5	19	14,5	11
I_C in mA	0.386	0.273	0.197	0.153	0.117	0.088

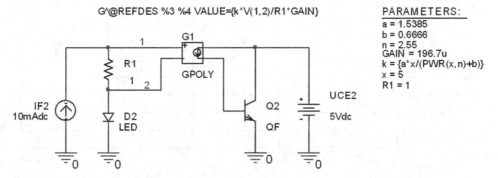

Fig. 3.48 Circuit for the dependence of the collector current on the reflector spacing x

a source GPOLY, see Fig. 3.48. Equation (3.32) can now be entered into this source at VALUE.

Analysis

- PSpice, Edit Simulation Profile
- Simulation Settings – Fig. 3.48: Analysis
- Analysis type: DC Sweep
- Options: Primary Sweep
- Sweep variable: Global Parameter
- Parameter Name: x
- Sweep type: Linear
- Start value: 0
- End value: 5
- Increment: 1 m
- Apply: OK
- PSpice, run

The characteristic curve in Fig. 3.49 is a useful approximation of the values in Table 3.12. If the current I_C (Q$_2$) is related to the maximum current $I_{Cmax} = 0.8046$ mA, then the

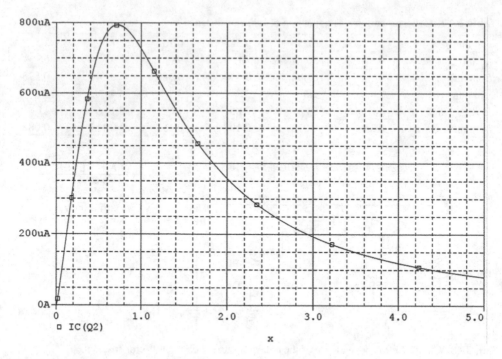

Fig. 3.49 Collector current as a function of reflector spacing x (in millimetres)

normalized representation according to Fig. 3.50 is obtained, which largely corresponds to the representation of the data sheet, see also Table 3.12.

Task: Proof of Function

The circuit shown in Fig. 3.51 can be used to demonstrate the operation of the reflective light barrier [14]. If a sufficiently large fraction of the light emitted by the LED is reflected onto the phototransistor, then its base-emitter voltage increases and so does the collector current and the voltage drop across the resistor R_3. The collector-emitter saturation voltage of transistor Q_1 assumes small values of 0.15–0.25 V and so the base-emitter voltage of Q_2 is not sufficient to drive the LED D_2. At higher values of R_3, smaller collector currents and thus higher spacing intervals Δx are sufficient to turn off the LED.

Analysis

- PSpice, Edit Simulation Profile
- Simulation Settings – Fig. 3.51: Analysis
- Analysis type: DC Sweep
- Options: Primary Sweep
- Sweep variable: Global Parameter
- Parameter Name: x

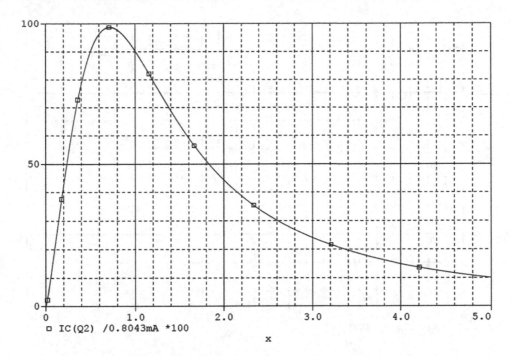

Fig. 3.50 $I_C (Q_2)/I_{Cmax}$ in percent as a function of reflector spacing x (in millimetres)

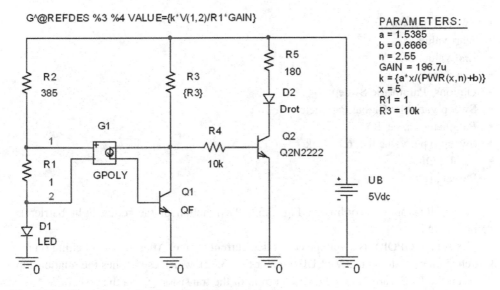

Fig. 3.51 Circuit for testing the reflex light barrier

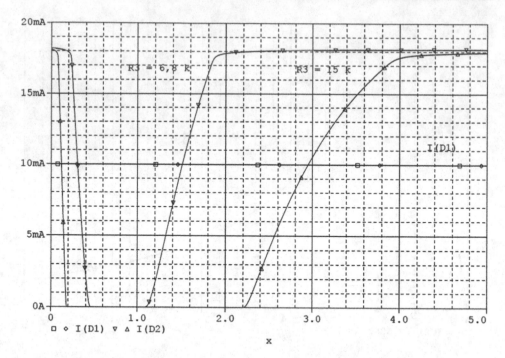

Fig. 3.52 Functional verification of the reflex light barrier with R_3 as parameter and x in mm

- Sweep type: Linear
- Start value: 0
- End value: 5
- Increment: 1 m
- Options: Parametric Sweep
- Sweep variable: Global Parameter
- Parameter Name: R3
- Sweep type: value list: 6.8–15 k
- Apply: OK
- PSpice, run

With the illustration according to Fig. 3.52, the function of the reflex light barrier is demonstrated.

The source GPOLY is a voltage controlled current source. After double-clicking on the switch symbol, you can use VALUE to enter Eq. (3.32), which establishes the relationship between the LED current and the base current of the transistor Q_1 via the resistor R_1.

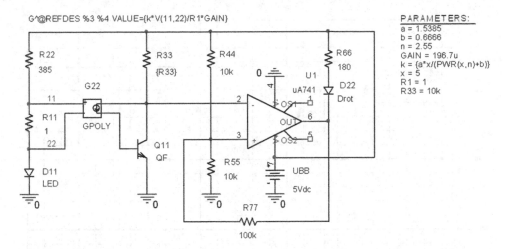

Fig. 3.53 Reflex light barrier with threshold switch

Task: Reflex Light Barrier with Threshold Switch

The circuit given in Fig. 3.53 according to [14] uses a threshold switch to set sharply defined reflector distance ranges in which the LED D_{22} is inactive. For this circuit, the resistor R_{33} is to be varied with values of 5 kΩ, 10 kΩ and 20 kΩ to capture the response of the LED.

Analysis

- PSpice, Edit Simulation Profile
- Simulation Settings – Fig. 3.53: Analysis
- Analysis type: DC Sweep
- Options: Primary Sweep
- Sweep variable: Global Parameter
- Parameter Name: x
- Sweep type: Linear
- Start value: 0
- End value: 5
- Increment: 1 m
- Options: Parametric Sweep
- Sweep variable: Global Parameter
- Parameter Name: R33
- Sweep type: value list: 5 k 10 k 20 k
- Apply: OK
- PSpice, run

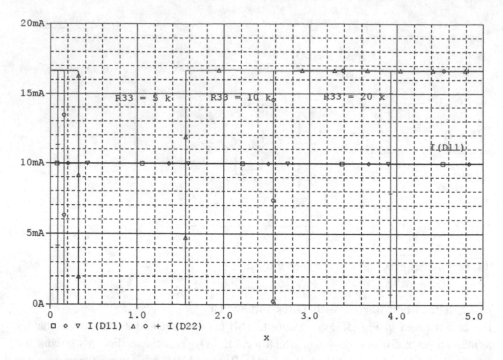

Fig. 3.54 Setting the switching thresholds using the resistor R_{33}

The analysis result according to Fig. 3.54 shows that the higher the value of the resistor R_{33}, the larger the range of the reflector distance over which the LED does not light up. Compared to the diagram in Fig. 3.51, the threshold switch results in steep switching edges.

3.10 Infrared Light Barrier

Infrared light barriers consist of a transmitter that emits IR light pulses to a receiver located opposite. If the light beam is interrupted, a warning signal is triggered at the receiver. Light barriers secure objects and operate counting devices. Delete rest up to and including point [15].

Circuit Description
The photoelectric sensor shown in Fig. 3.55 shows the transmitter constructed with the 555D timer circuit and the GaAs IR diode LD 271, and the receiver containing the npn phototransistor SFH 309, the operational amplifier µA 741, and the rectifier stage with diodes 1 N 4148 [14]. When light is interrupted, transistor Q_3 is turned on. As a result, the buzzer circuit comes to operating voltage and the three-electrode buzzer KPEG 132 emits a

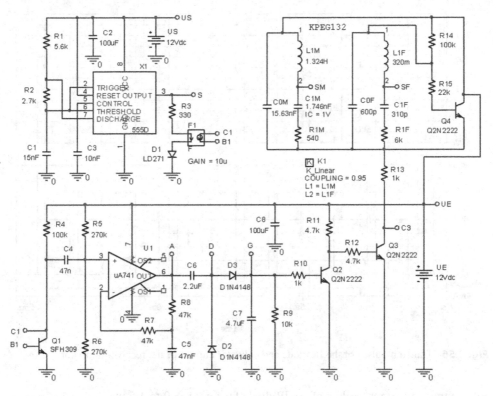

Fig. 3.55 Circuit of the infrared light barrier

loud signal. The transmission of the pulses on the part of the transmitter diode LD 271 is carried out in simulation with a current-controlled current source F from the analog library. The maximum range of this circuit of five meters corresponds to a value GAIN = 5u = 5E-06 = $5–10^{-6}$. Thus, only a fraction of the IR transmit pulse I_F as current I_B to the base of the phototransistor.

It is GAIN = I_B/I_F. The semiconductor devices are modeled as follows:

.model LD271 D IS = 11.27p N = 2.158 RS = 0.63 EG = 1.6

. model SFH309 NPN IS = 10f BF = 500 VAF = 100 IKF = 0.2 ISE = 200p NE = 3 TF = 1.5n.

The buzzer circuit, which is located between nodes C_3 and U_E, corresponds to the specifications of the manufacturer Kingstate Electronics Corporation [15] for the self-triggering piezoelectric buzzer type KPEG 132, see Sect. 8.2.

Task: Transmission of IR Light Pulses

In the circuit shown in Fig. 3.55, the value GAIN = 10 u is to be entered. In the time range of a few milliseconds are to be analysed and displayed:

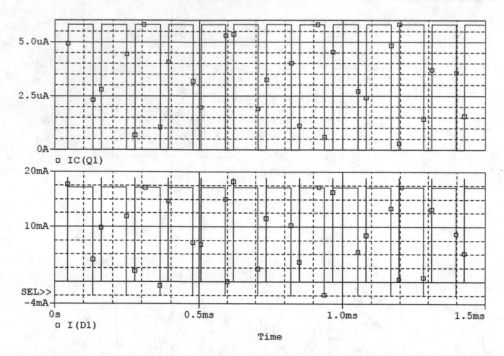

Fig. 3.56 Transmit pulses of the IR diode and collector current of the receiver transistor

- The transmit current pulses of the IR diode D_1 for $\Delta t = 0$ to 1.5 ms
- The collector current of the receiver transistor Q_1 for $\Delta t = 0$ to 1.5 ms
- The stresses at nodes D and G for $\Delta t = 0$ to 4 ms
- The voltages V(SM) at the main segment of the buzzer for $\Delta t = 0$ to 25 ms

Analysis

- PSpice, Edit Simulation Profile
- Simulation Settings – Fig. 3.55: Analysis
- Analysis type: Time Domain (Transient)
- Options: General Settings
- Run to time: 1.5 ms, 4 ms, 30 ms
- Start saving data after: 0
- Maximum step size: 1 us
- Apply: OK
- PSpice, run

The analysis result of Fig. 3.56 shows the transmit pulses of the photodiode with the switching frequency $f_p \approx 8.7$ kHz and the collector current pulses of the phototransistor. A fraction of the current of the transmitting diode D_1 reaches the base of the phototransistor as

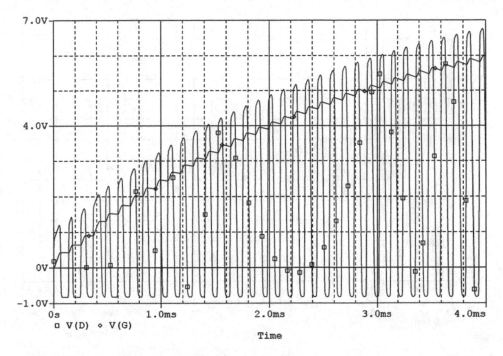

Fig. 3.57 Time structure of the stresses at nodes D and G

I_B (Q_1). For the collector current of the phototransistor, I_C (Q_1) = I_B (Q_1)–B_N. Where B_N is the large-signal current gain in normal operation.

At node D the amplified AC voltage is present and at node G the time characteristic of the rectified voltage appears, see Fig. 3.57.

The voltage difference V(U_E) – V(C_3) is calculated with the short decay time of $t \approx 30$ ms. The buzzer circuit thus has no operating voltage. Thus, for the considered case where the IR light pulses are transmitted, the buzzer KPEG 132 remains silent. The proof of this is provided by Fig. 3.58.

Task: Interruption of the IR Light Pulses

In the circuit according to Fig. 3.55, the value GAIN = 1 f must now be entered. This means that practically no IR transmission pulses reach the receiver. Using the analysis steps of the previous task, the buzzer oscillations V(SM) and V(SF) are to be analyzed and displayed in the time range $\Delta t = 0$ to 3 ms:

When the light is interrupted, the voltage at the collector of transistor Q_3 drops to the low level of its collector-emitter saturation voltage, so that the buzzer circuit is supplied with the nearly full operating voltage of $U_E = 12$ V. The buzzer voltages appear as stable oscillations. The oscillations generated by the main segment M (MAIN) have a higher amplitude than those of the feedback segment F (FEEDBACK), see Fig. 3.59.

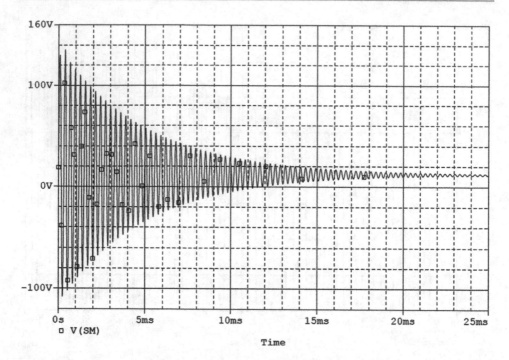

Fig. 3.58 Decaying oscillations at the main segment of the buzzer

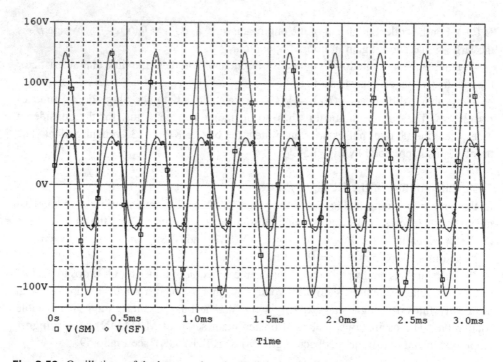

Fig. 3.59 Oscillations of the buzzer when the IR light barrier is interrupted

References

1. Perkin Elmer optoelectronics: Datenblatt des Fototransistors A1060. Wiesbaden
2. Wirsum, S.: Das Sensor-Kochbuch. IWT, Bonn (1994)
3. Faulhaber GmbH: DC Kleinmotoren. Technische Informationen, Schöneich (2009)
4. Baumann, P.: Sensorschaltungen. Vieweg+Teubner, Wiesbaden (2010)
5. Justus, O.: Dynamisches Verhalten elektrischer Maschinen. Vieweg+Teubner, Wiesbaden (1993)
6. Siemens: Datenblatt zur Fotodiode BFX 90 (1997)
7. Texas Instruments: Datenblatt zum Licht-Spannungswandler TSL 250 R. Dallas (2001)
8. HAMAMATSU: Datenblatt zum RGB-Farbsensor S7505-01
9. Siemens: Datenblatt zum Fototransistor BP 103 (1995)
10. Federau, J.: Operationsverstärker. Vieweg+Teubner, Wiesbaden (1998)
11. OMRON: Datenblatt zum Gabelkoppler EE-SX 1106. München (1999)
12. OSRAM: Datenblatt zur Reflexlichtschranke SFH 9201 (1999)
13. Abrams, E., Trabula, Y., Habben, M., Nana, C.: Reflexlichtschranke. Projektarbeit Hochschule, Bremen (2014)
14. Härtl, A.: Optoelektronik in der Praxis. Härtl, Hirschau (1998)
15. Kingstate Electronics Corp.: Datenblatt zum Summer KPEG132. Taipei (2016)

Force Sensors

4

4.1 Definition of Force

If a force F imparts an acceleration $a = 1$ m/s^2 to a body of mass m $= 1$ kg, then it has the magnitude of 1 Newton, 1 N $= 1$ kgm/s^2. It is according to Eq. (4.1)

$$F = m \cdot a \qquad (4.1)$$

The following applies specifically to the weight force F_G according to Eq. (4.2)

$$F_G = m \cdot g \qquad (4.2)$$

with the acceleration due to gravity $g = 9.81$ m/s^2.

Example: If a body with the mass $m = 100$ g (100 grams) falls to earth, then a weight force $F_G = m - g = 100$ g-9.81 m/s$^2 = 0.981$ N acts.

4.2 Foil Force Sensor

Figure 4.1 shows the structure of the force sensor FSR 400 (Force Sensing Resistor) and a circuit for simulating the dependence of the sensor resistance on the mass.

A semiconducting polymer layer is applied to a carrier film. A double-sided adhesive layer arranged at the edges connects this lower FSR carrier foil with an upper carrier foil and at the same time serves as a spacer between the two foils. On the underside of the upper carrier foil, finger-like interlocking electrodes are printed. If a mass in the form of an impact

P. Baumann, *Selected Sensor Circuits*,
https://doi.org/10.1007/978-3-658-38212-4_4

Fig. 4.1 Structure of the FSR
400 force sensor and circuit for
the sensor characteristic curve

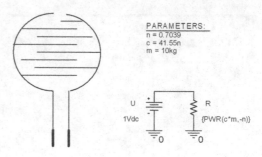

load acts on the active surface, resistance bridges are closed, which greatly reduces the
sensor resistance [1].

Data Sheet

Force sensor FSR 400 from Interlink Electronics [1], diameter: 7.62 mm, length: 38 mm,
thickness: 200–750 µm. Switch-on force: 0.2–1 N, nominal force: 100 N (maximum value
near saturation) $R > 1$ MΩ (unloaded), $I_{max} = 1$ mA, $U = 1$ V to 5 V, *Ptotmax* $= 1$ mW,
$TK_R = -0.8\%/K$, mechanical response time <2 ms, electrical response time $= 0.1$–10 ms,
lifetime >10–10^6 switching cycles, maximum relative humidity: 85%. Force sensors are
used for touch control of electronic devices, for games, remote controls, navigation
electronics and in medical technology.

In the manufacturer's data sheet, the characteristic curve $R = \mathrm{f}(m)$ of the FSR 400 sensor
for the range $m = 50$ g to 10 kg appears approximately as a straight line on a double
logarithmic scale. This characteristic curve can be described with Eq. (4.3).

$$R = (c \cdot m)^{-n} \tag{4.3}$$

The exponent n follows from logarithmization with Eq. (4.4).

$$n = \frac{\lg(R_2/R_1)}{\lg(m_1/m_2)} \tag{4.4}$$

From the manufacturer's characteristic curve, the following pairs of values are evaluated
using Eqs. (4.3 and 4.4): (1) $R_1 = 240$ Ω at $m_1 = 10$ kg and (2) $R_2 = 10$ kΩ at $m_2 = 50$ g.
We obtain $n = 0.7039$ and $c = 41.55$–10^{-9}.

Task

The dependence of the sensor resistance R *on* the short-time acting mass m can be
represented in the range from 50 g up to 10 kg with the circuit shown in Fig. 4.1.

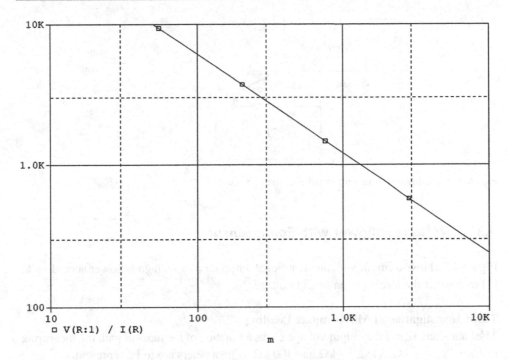

Fig. 4.2 Simulated decrease of sensor resistance with increase of mass

Analysis

- PSpice, Edit Simulation Profile
- Simulation settings – Fig. 4.1: Analysis
- Analysis type: DC Sweep
- Options: Primary Sweep
- Sweep variable: Global Parameter
- Parameter Name: m
- Sweep type: Logarithmic, Decade
- Start value: 50
- End value: 10 k
- Points/Decade: 100
- Takeover: OK
- PSpice, run

The analysis result according to Fig. 4.2 largely agrees with the manufacturer's specifications. Equation (4.2) applies to the weight force F_G.

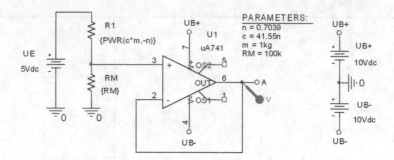

Fig. 4.3 Dependence of the output voltage on the ground

4.3   Voltage Follower with Force Sensor

Figure 4.3 shows a circuit with an operational amplifier as a voltage follower according to [1] to convert the force into an output voltage.

Task: Investigation of Mass-Impact Loading
The characteristics of the output voltage U_A as a function of the mass m with the measuring resistor $R_M = 1$ kΩ, 3 kΩ, 10 kΩ and 100 kΩ as parameters are to be represented.

Analysis

- PSpice, Edit Simulation Profile
- Simulation settings – Fig. 4.3: Analysis
- Analysis type: DC Sweep
- Options: Primary Sweep
- Sweep variable: Global Parameter
- Parameter Name: m
- Sweep type: Linear
- Start value: 20
- End value: 1 k
- Increment: 50 m
- Options: Parametric Sweep
- Sweep variable: Global Parameter
- Parameter Name: RM
- Sweep type: Value list: 1 k 3 k 10 k 100 k
- Apply, OK
- PSpice, run

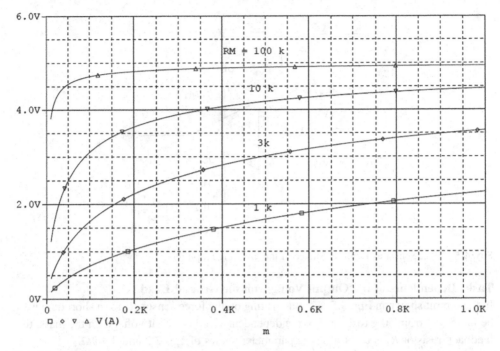

Fig. 4.4 Effect of ground on the output voltage of the voltage follower

The analysis result according to Fig. 4.4 shows that the output voltage increases to a certain degree with increasing mass or weight force. The voltage follower provides a high input resistance.

4.4 Inverting Amplifier with Force Sensor

In the circuit shown in Fig. 4.5, the force sensor is a component of an inverting amplifier [2]. The output voltage of this amplifier is detected with positive input voltage with Eq. (4.5).

$$U_A = -\frac{R_2}{R_1} \cdot U_E \tag{4.5}$$

If the negative pole of the input voltage source is connected to the resistor R_1 as shown in Fig. 4.5, *positive* values are obtained for the output voltage.

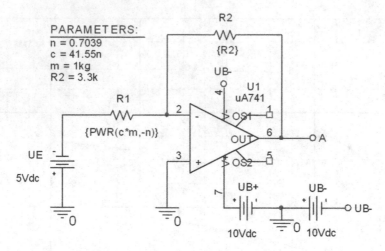

Fig. 4.5 Force sensor as input resistance of the inverting amplifier

Task: Dependence of the Output Voltage on the Ground Load

In the circuit shown in Fig. 4.5, the mass acting on the force sensor R_1 for a short time is to be increased from 50 g to 1 kg. For the dependence of the output voltage on the mass, the feedback resistor R_2 is to assume the parameter values of 1.5, 2.2 and 3.3 kΩ.

Analysis

- PSpice, Edit Simulation Profile
- Simulation settings – Fig. 4.5: Analysis
- Analysis type: DC Sweep
- Options: Primary Sweep
- Sweep variable: Global Parameter
- Parameter Name: m
- Sweep type: Linear
- Start value: 50
- End value: 1 k
- Increment: 50 m
- Options: Parametric Sweep
- Sweep variable: Global Parameter
- Parameter Name: R2
- Sweep type: Value list: 1.5 k, 2.2 k, 3.3 k
- Apply, OK
- PSpice, run

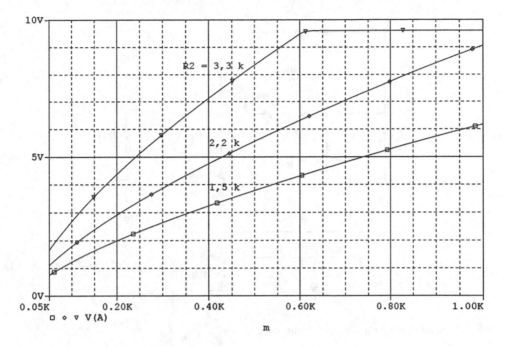

Fig. 4.6 Output voltage as a function of ground

The analysis result according to Fig. 4.6 shows that the output voltage increases at higher ground loads of the sensor due to the associated reduction of its resistance until the saturation limit of the operational amplifier is reached.

4.5   Schmitt Trigger Multivibrator with Force Sensor

In the circuit shown in Fig. 4.7, a force sensor located in the feedback branch of the multivibrator is used to control the oscillation frequency. The period T is calculated according to [3] for a duty cycle $v = t_p/T = 1{:}2$ with Eq. (4.6).

$$T = 2 \cdot R_{FSR} \cdot C \cdot \ln\left(1 + 2 \cdot R_1/R_2\right) \qquad (4.6)$$

Task: Verification of Vibrations
For a short-term load on the force sensor with a mass of 100 g or 3 kg, the oscillations at the output of the multivibrator are to be simulated for the period from 0 to 3.6 ms.

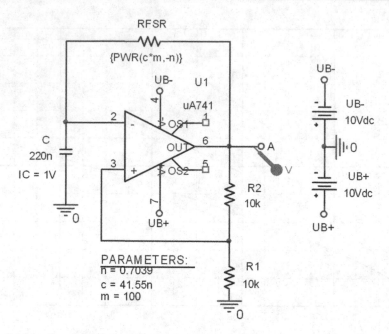

Fig. 4.7 Schmitt trigger multivibrator with force sensor

Analysis

- PSpice, Edit Simulation Profile
- Simulation settings – Fig. 4.7: Analysis
- Analysis type: Time Domain (Transient)
- Options: Primary Sweep
- Run to time: 3.6 ms
- Start saving data after: 0 s
- Transient Options
- Maximum step size: 1 us
- Options: Parametric Sweep
- Sweep variable: Global Parameter
- Parameter Name: m
- Sweep type: Value list: 100, 3 kg
- Apply, OK
- PSpice, run

Figures 4.8 and 4.9 show the influence of the load on the force sensor. From Eq. (4.3), we obtain $R_{FSR} = 6137\ \Omega$ for $m = 100$ g. Here follows $T = 2.97$ ms. In contrast, $R_{FSR} = 560\ \Omega$ for $m = 3$ kg with $T = 0.27$ ms.

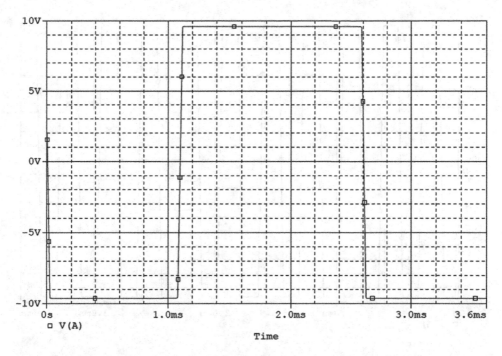

Fig. 4.8 Multivibrator vibrations at a mass load of one hundred grams

4.6   Schmitt Trigger with Force Sensor

In the circuit shown in Fig. 4.10, a non-inverting trigger is controlled via a force sensor [2]. The zero point of the hysteresis is shifted with the voltage applied to the N input [3, 4]. In the example, $U_N = 2.5$ V. If only the positive operating voltage with $U_B = 5$ V is applied to the LM 324 operational amplifier, the following maximum and minimum values of the output voltage are obtained: $U_{S+} = 4.11$ V and $U_{S-} = 36.6$ mV [5]. The switch-on trigger threshold is obtained with Eq. (4.7) as follows

$$U_{Eein} = U_N \cdot \left(1 + \frac{R_1}{R_2} \right) \tag{4.7}$$

Neglecting U_{S-} according to [4], the turn-off trigger threshold is captured by Eq. (4.8).

$$U_{Eaus} = U_N \cdot \left(1 + \frac{R_1}{R_2} \right) - U_{S+} \cdot \frac{R_1}{R_2} \tag{4.8}$$

From Eqs. (4.7) and (4.8), the switching hysteresis is given by Eq. (4.9).

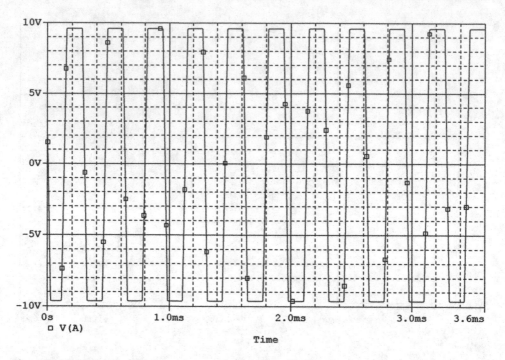

Fig. 4.9 Multivibrator vibrations at a mass load of three kilograms

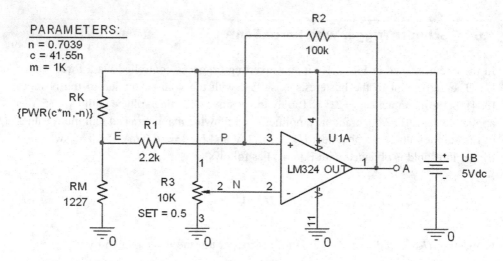

Fig. 4.10 Controlling the Schmitt trigger via the force sensor

$$U_H = U_{S+} \cdot \frac{R_1}{R_2} \tag{4.9}$$

You get $U_H = U_{Eein} - U_{Eaus} = 2.55 \text{ V} - 2.465 \text{ V} = 0.09 \text{ V}$.

Task: Verification of Hysteresis

The force sensor R_K is to be subjected for a short time to (an impact with) a mass which is increased from m = 50 g to 2 kg and then reduced from m = 2 kg to 50 g. The force sensor R is to be switched off at $m = 2$ kg. Output A is to be switched on at $m_{ein} = 1037$ g from $U_{S-} = 36.6$ mV to $U_{S+} = 4.11$ V and switched off at $m_{aus} = 937$ g.

Solution:

According to Eq. (4.3), we first obtain $R_{Kaus} = 1270.52 \ \Omega$ for $m_{aus} = 937$ g and $R_{Kein} = 1183 \ \Omega$ for $m_{ein} = 1037$ g. For the measurement resistance, the arithmetic mean of the R_K values is used with $R_M = 1227 \ \Omega$ [5]. From the voltage divider at the input, the turn-on voltage is obtained by Eq. (4.10).

$$U_{Eein} = \frac{R_M}{R_M + R_{Kein}} \cdot U_B \tag{4.10}$$

The turn-off voltage follows with Eq. (4.11) to

$$U_{Eaus} = \frac{R_M}{R_M + R_{Kaus}} \cdot U_B \tag{4.11}$$

The calculation yields $U_{Eein} = 2.5440$ V and $U_{Eaus} = 2.4544$ V. Thus, the switching hysteresis reaches the value $U_H = U_{Eein} - U_{Eaus} = 89.6$ mV. With the specification of $R_1 = 2.2$ kΩ, we obtain the positive feedback resistor from the conversion of Eq. (4.9) $R_2 = 100.92$ kΩ. According to the E 6 series, $R_2 = 100$ kΩ is selected.

Analysis

- PSpice, Edit Simulation Profile
- Simulation settings – Fig. 4.10: Analysis
- Analysis type: DC Sweep
- Options: Primary Sweep
- Sweep variable: Global Parameter
- Parameter Name: m
- Sweep type: Linear
- Start value: 50
- End value: 2 k
- Increment: 0.5
- Apply, OK
- PSpice, run

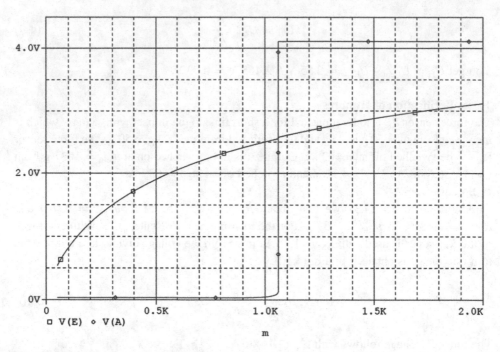

Fig. 4.11 Input and output voltage as a function of increasing ground

The analysis result according to Fig. 4.11 shows the dependence of the input and output voltage on the increasing ground. Switching on takes place at $m = 1059.5$ g. If the mass acting on the sensor is reduced, then switching off occurs at $m = 940.5$ g according to Fig. 4.12. A symmetry with $m = 1$ kg +/− 59.5 g.In the transfer characteristic shown in Fig. 4.13, switch-on occurs at $U_{\text{E ein}} = 2.564$ V. Switching off occurs at $U_{\text{Eaus}} = 2.459$ V, see Fig. 4.14.

4.7   Counting Circuit

The counting circuit shown in Fig. 4.15 serves as a preliminary stage for setting up an arrangement with which the number of pressure pulses applied to the force sensor can be recorded. The TTL JK flip-flop 7490A is driven with clock T by the digital signal source DSTIM1 and switches with the HL edge. To realize the states from 0 to 9, the input CKB must be connected to the output QA of the first flip-flop. The inputs R_{01} and R_{02} receive an initial pulse from the source DSTIM2, which sets the counter to a defined initial state [5]. The signals of the JK flip-flop reach the TTL decoder 7448, from which a seven-segment display with a common cathode can be controlled.

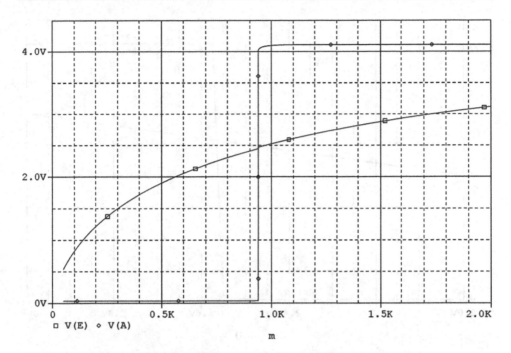

Fig. 4.12 Input and output voltage as a function of decreasing ground

Task: Display of the Pulse Diagram

With the circuit according to Fig. 4.15 the pulse diagram for the pulse R, the clock signal T and for the output signals of the counter as well as the decoder is to be displayed. The analysis is to be carried out from zero up to 16 s.

Analysis

- PSpice, Edit Simulation Profile
- Simulation settings – Fig. 4.15: Analysis
- Analysis type: Time Domain (Transient)
- Options: General Settings
- Run to time: 16 s
- Start saving data after: 0 s
- Transient Options
- Maximum step size: 10 ms
- Apply, OK
- PSpice, run

With each one clock period, a pulse appears at output QA of counter7490A, two clock periods result in a pulse at QB, four clock periods generate a pulse at QC, and at the HL

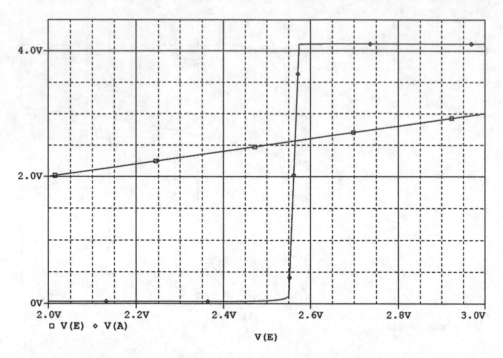

Fig. 4.13 Transfer characteristic for the switch-on process

edge of the eighth clock period, the counter is reset. The 8-4-2-1 code can be used to encode the decimal digits 1–9, see Table 4.1 after Eq. (4.6).

The output voltages OA to OG of the decoder 7448 refer to the seven-segment display. The pulse diagram for the circuit according to Fig. 4.15 is shown in Fig. 4.16.

4.8 Counting the Impact Load Pulses

The circuit shown in Fig. 4.17 presents a possible application of the force sensor, with which small parts that hit the FSR resistor layer during the production process can be recorded numerically from 0 to 9. Starting from the bounce-free circuit of the Schmitt trigger shown in Fig. 4.10, the voltage divider is pulse-controlled with the associated force sensor R_K. An impinging mass of $m = 1.1$ kg causes the resistance R_K to turn out smaller than the measuring resistance R_M, whereby the non-inverting Schmitt trigger reaches the potential of its positive saturation voltage at its output. The clock T follows the pulse voltage U_P and is applied to pin 14 of the JK flip-flop. The output voltages of the decoder are connected to the series resistors of the segment LED.

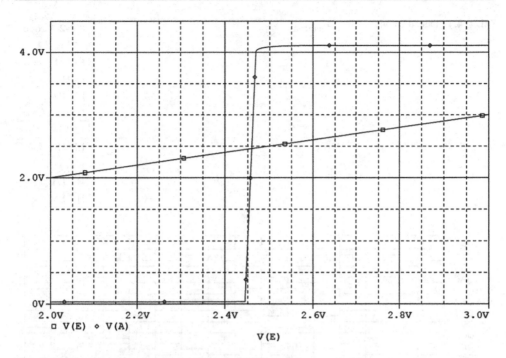

Fig. 4.14 Transfer characteristic for the switch-off process

COMMAND1 = 0s 0
COMMAND2 = LABEL=STARTLOOP
COMMAND3 = 0.5s 1
COMMAND4 = 1s 0
COMMAND5 = 1.5s GOTO STARTLOOP -1 TIMES
TIMESTEP = 0.5s

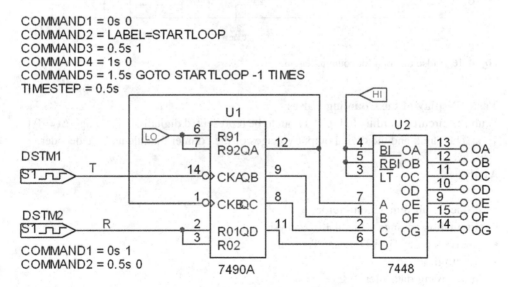

Fig. 4.15 Circuit with decimal counter and seven-segment decoder

Table 4.1 Decadic code

Decimal	8-4-2-1 code
0	LLLL
1	LLLH
2	LLHL
3	LLHH
4	LHLL
5	LHLH
6	LH HL
7	L H HH
8	H LL L
9	H LLH

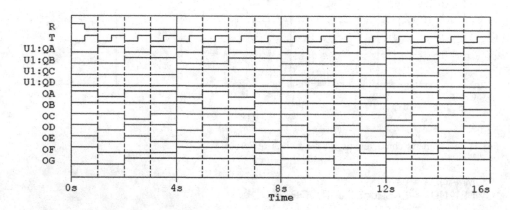

Fig. 4.16 Pulse diagram for counting circuit

Task: Display of the Counting Pulses

With the circuit according to Fig. 4.17 are to be represented digitally in the range $\Delta t = 0$ to 16 s: The clock and the digital output voltages of the counter module and the decoder.

Analysis

- PSpice, Edit Simulation Profile
- Simulation settings – Fig. 4.17: Analysis
- Analysis type: Time Domain (Transient)
- Run to time: 16 s
- Start saving data after: 0 s
- Maximum step size: 10 ms
- Apply, OK
- PSpice, run

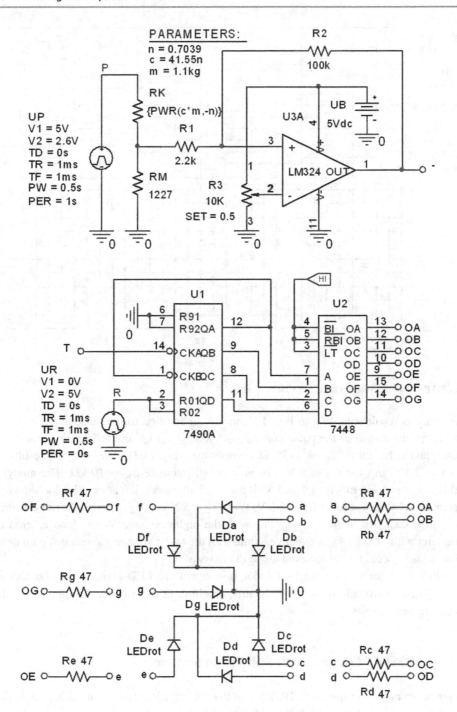

Fig. 4.17 Counter with seven-segment display for indication of shock loads

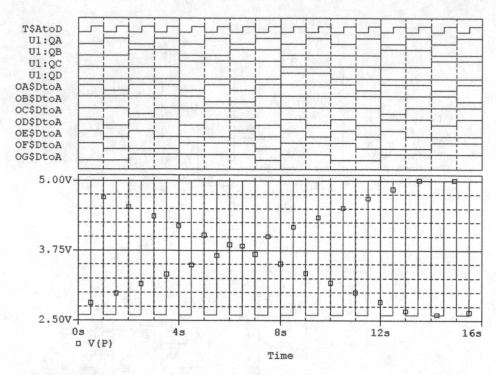

Fig. 4.18 Digital voltages and input pulse voltage

The analysis result according to Fig. 4.18 shows, in addition to the pulse diagram (as in Fig. 4.16), the course of the pulse voltage U_P. At $m = 1.1$ kg, the resistance of the force sensor takes the value $R_K = 1135\ \Omega$ according to Eq. (4.3). The analysis results in $I(R_K) = 2.109$ mA for $U_P = 5$ V. For $m = 50$ g, calculate $R_K = 10$ kΩ. The analysis yields $I(R_K) = 0.446$ mA for $U_P = 5$ V. If the incident mass is left at $m = 1.1$ kg, then this current of 0.446 mA is set at $U_P = 2.6$ V. With $V_1 = 5$ V and $V_2 = 2.6$ V, the pulse source U_P is thus designed in such a way that with the higher voltage V_1 the load entered as parameter with $m = 1.1$ kg is effective, while with the lower voltage V_2 with this parameter value a load occurs which corresponds to the mass of 50 g.

In Fig. 4.19, for the time range of 4–5 s, it is shown that LED segments D_b, D_c, D_f and D_g carry a forward current of about 10 mA, resulting in the number 4 appearing in the seven-segment display.

4.9   MOSFET Control with Foil Force Sensor

The forward current of the GaAs IR diode of the A4N25 optocoupler can be adjusted with the circuit shown in Fig. 4.20. A MOSFET or a bipolar transistor is driven via the voltage divider with the force sensor FSR [6]. In this case, the FSR current remains far below its

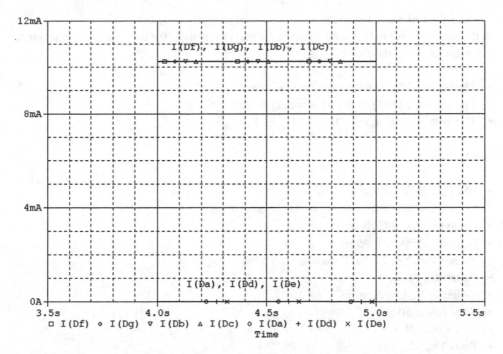

Fig. 4.19 LED segment currents to represent the number 4

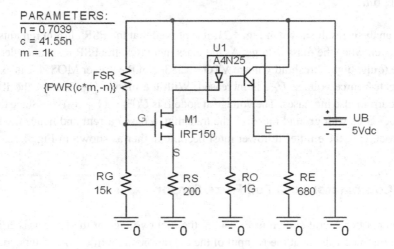

Fig. 4.20 MOSFET control via the FSR voltage divider

limit value. As the mass m increases, the resistance of the FSR decreases which allows the threshold voltage of the n-channel enhancement MOSFET to be overcome. A small portion of the LED transmit current is optically transmitted to the collector-base path of the Si-npn receiver transistor at whose emitter the voltage drop across resistor R_E is evaluated.

Task: Effect of Mass Increase

In the circuit shown in Fig. 4.20, increase the mass from $m = 50$ g to 10 kg. As a function of the mass are to be shown:

- The current I(FSR) of the foil force sensor
- The current I(D1:1) of the transmitting diode
- The voltage drop at node E

Analysis

- PSPICE, Edit Simulation Profile
- Simulation Settings – Fig. 4.20: Analysis
- Analysis type: DC Sweep
- Options: Primary Sweep
- Sweep variable: Global Parameter
- Parameter Name: m
- Sweep type: Logarithmic
- Start value: 50
- End value: 10 k
- Point/Decade: 100
- Apply: OK
- PSpice, run

With the analysis result shown in Fig. 4.21, it is proved that the FSR current remains below the limit even when the mass is large. As the mass increases, the FSR resistance decreases. Thus, the (quite high) threshold voltage VTO = 2.83 V of the power MOSFET is exceeded and the gate-source voltage U_{GS} is increased. With the higher U_{GS} values, the IR diode transmit current also increases. The voltage at node E is $U(E) = (1 + B_N)–I_B$. Since the LED current increases with ground increase, the transmitted base current and hence the voltage drop at node E of the emitter follower must become higher as shown in Fig. 4.22.

4.10   Comparator with Foil Force Sensor

In the comparator circuit shown in Fig. 4.23, the input voltage at the P input is compared with the reference voltage at the N input of the operational amplifier. In the unloaded state of the foil force sensor, the red LED lights up because $R_S > R_M$ and thus $U_P < U_N = 2$ V. When the sensor with mass $m > 1$ kg is tapped, the output voltage assumes the value of the positive saturation voltage $U_S = 4.11$ V and the green LED lights up instead of the red LED.

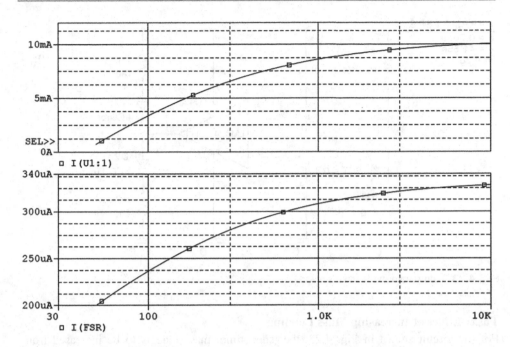

Fig. 4.21 Force sensor current and transmitter diode current as a function of mass

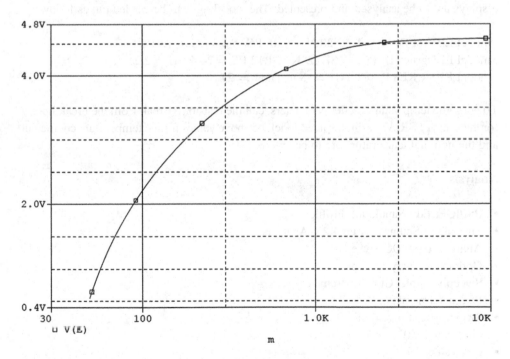

Fig. 4.22 Dependence of the output voltage on the ground

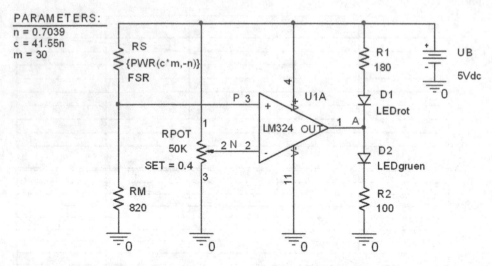

Fig. 4.23 Comparator with force sensor

Task: Effect of Increasing Mass Loading

For the circuit shown in Fig. 4.23, the (short-time) mass load is to be increased from $m = 30$ g to 10 kg. The dependencies $U(P)$, $U(N)$, $U(A) = f(m)$ and the resulting LED displays are to be analysed and presented. The modeling is to be carried out as follows:

. model LEDrot D (IS = 1.2E-20 N = 1.46 RS = 2.4 EG = 1.9)
. model LEDgruen D (IS = 9.8E-29 N = 1.12 RS = 24.4 EG = 2.2)
. model RFSR RES R = 1 TC1 = -8 m Tnom = 27.

To enter the temperature model parameters, call the resistor RBreak from the break library for the sensor FSR. Via Edit, PSpice Model the above values for the temperature coefficient and the nominal temperature are to be entered.

Analysis

- PSPICE, Edit Simulation Profile
- Simulation Settings – Fig. 4.23: Analysis
- Analysis type: DC Sweep
- Options: Primary Sweep
- Sweep variable: Global Parameter
- Parameter name: m
- Sweep type: Logarithmic
- Start value: 50
- End value: 10 k

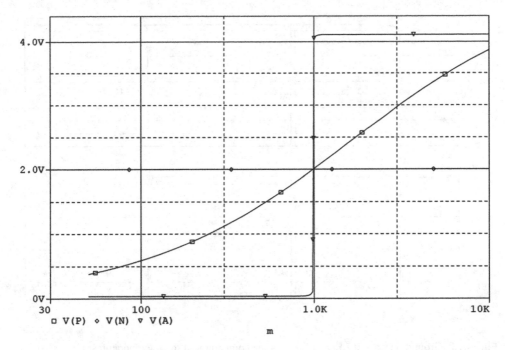

Fig. 4.24 Switching point of the output voltage

- Point/Decade: 100
- Apply: OK
- PSpice, run

The analysis result according to Fig. 4.24 shows the constant reference voltage $U_N = 2$ V. The switching point lies at the intersection of the voltages at the N and P inputs. The short-time mass load with $m = 1$ kg corresponds approximately to tapping the foil force sensor with a finger.

Task: Effect of FSR Temperature Dependence
The negative temperature coefficient TC1 $= -0.8\%$/K of the sensor specified by the manufacturer in [1] was previously taken into account in the modelling. The effect of this temperature dependence on the onset of the LED currents as a function of the mass is to be analyzed. The analysis is to be carried out for temp $= -40$, 27 and 60 °C.

Analysis

- PSPICE, Edit Simulation Profile
- Simulation settings – Fig. 4.23: Analysis
- Analysis type: DC Sweep

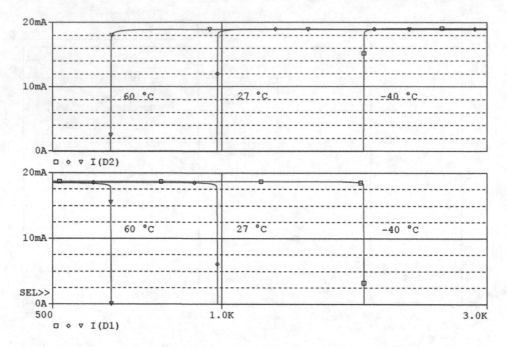

Fig. 4.25 Ground-dependent LED displays on the comparator at three temperatures

- Options: Primary Sweep
- Sweep variable: Global Parameter
- Parameter name: m
- Sweep type: Logarithmic
- Start value: 50
- End value: 10 k
- Point/Decade: 100
- Options: Parametric Sweep
- Sweep variable: Temperature
- Sweep type: Value List: −40, 27, 60
- Apply: OK
- PSpice, run

The previously explained curve of the LED currents as a function of the mass at three temperatures is shown in Fig. 4.25. In the operating temperature range from −40 to +85 °C, the LED currents start at quite different mass effects.

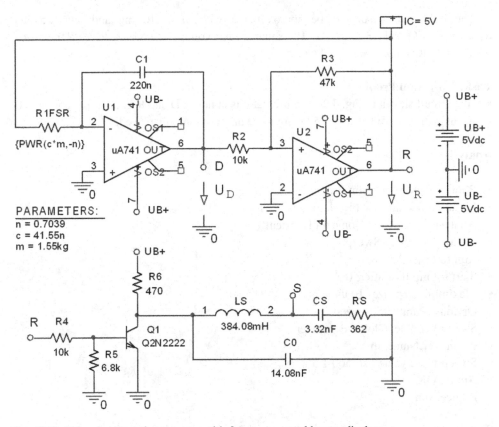

Fig. 4.26 Triangle-rectangle generator with force sensor and buzzer display

4.11   Triangle-Rectangle Generator with Foil Force Sensor

The circuit shown in Fig. 4.26 contains in the first stage an integrator whose behaviour is determined by the time constant $\tau = R_1 - C_1$ and in the second stage a Schmitt trigger whose output voltage U_R is fed back to the integrator input. At the integrator output, the triangle voltage with the amplitude U_D is formed. The following applies: $U_D/U_R = R_2/R_1$. The switching frequency of the generator is obtained according to [7] as follows

$$f = \frac{R_3}{4 \cdot R_1 \cdot R_2 \cdot C_1} \tag{4.12}$$

The resistance R_1 in the above equation corresponds to the foil force sensor R_{1FSR} which can be influenced by the weight force F_G. For a (short-time) mass load with $m = 1.5$ kg, Eq. (4.3) yields the value $R_{1FSR} = 912\ \Omega$ and thus $f = 5.86$ kHz according to Eq. (4.12).

The unloaded sensor can be approximated with $m = 10$ mg and then obtains $R_{1FSR} = 4$ MΩ and $f = 1.34$ Hz. The square-wave voltage at node R drives a transistor whose output is a piezoelectric buzzer.

Task: Vibration Proof
For the circuit shown in Fig. 4.26, the oscillations at nodes D and R in the time range $t = 0$ to 10 ms with $m = 500$ g and 1.5 kg are to be analyzed and displayed.

Analysis

- PSpice, Edit Simulation Profile
- Simulation settings – Fig. 4.26: Analysis
- Analysis type: Time Domain (Transient)
- Options: Primary Sweep
- Run to time: 2.5 ms
- Start saving data after: 0 s
- Maximum step size: 10 us
- Options: Parametric sweep
- Sweep variable: Global Parameter
- Parameter Name: m
- Sweep type: value list: 500 1.5 k
- Apply, OK
- PSpice, run

With the different weight forces applied to the foil force sensor, the switching frequencies can be changed, see Fig. 4.27.

With a higher mass effect, the sensor resistance and the period duration decrease, with which the frequency increases.

Task: Buzzer Display
The circuit shown in Fig. 4.26 contains at the generator output R a circuit with a piezoelectric buzzer of the type EPZ-27MS47, see Chap. 8. The sinusoidal buzzer oscillations are detected at the node S.

The mass incident on the force sensor shall be varied as $m = 10$ mg, 1.5 kg and 2 kg.

The resulting oscillation amplitudes of the buzzer are to be analyzed in the time range $\Delta t = 0$ to 5 ms.

Analysis

- PSpice, Edit Simulation Profile
- Simulation settings – Fig. 4.26: Analysis
- Analysis type: Time Domain (Transient)

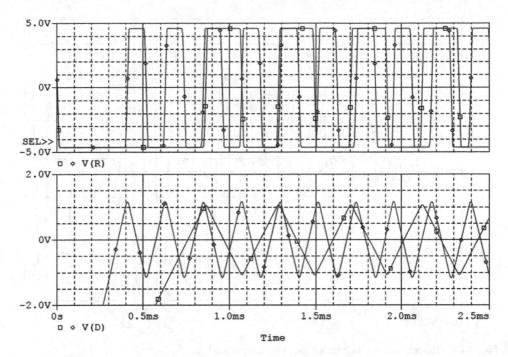

Fig. 4.27 Generator oscillations for different mass loading of the sensor

- Options: Primary Sweep
- Run to time: 5 ms
- Start saving data after: 0 s
- Maximum step size: 10 us
- Options: Parametric sweep
- Sweep variable. Global parameter
- Parameter name: m
- Sweep type: value list: 10 m 1.5 k
- Apply, OK
- PSpice, run

In Fig. 4.28 it can be seen that no vibration is detectable for the unloaded sensor after a short time. This case is simulated with the mass $m = 10$ mg. In contrast, oscillations occur for $m = 1.5$ kg.

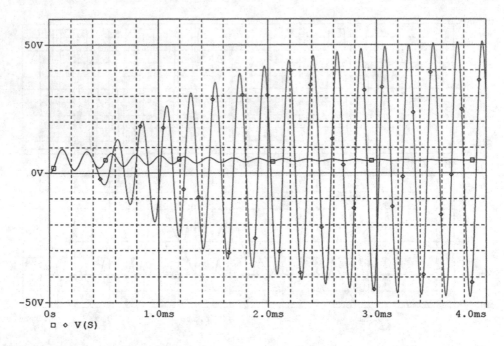

Fig. 4.28 Buzzer oscillations without and with a mass load

References

1. Interlink Electronics: Datenblatt FSR 400, Ausgabe 9 (2000)
2. Interlink Electronics: FSR Integration Guide and Evaluation Parts Catalog (2010)
3. Weddigen, C., Jüngst, W.: Elektronik. Springer, Berlin (1993)
4. Böhmer, E., Ehrhardt, D., Oberschelp, W.: Elemente der angewandten Elektronik. Vieweg +Teubner, Wiesbaden (2010)
5. Gersdorf, J.-P., Himmel, N., Schlemm, P., Schmitt, N.: Kraftsensor. Projektarbeit Hochschule, Bremen (2015)
6. Sensitronics: Force Sensing Resistor Theory and Applications. Firmenschrift (2015)
7. Koß, G., Reinhold, W.: Lehr- und Übungsbuch Elektronik. Fachbuchverlag, Leipzig (1998)

Pressure Sensors

5

5.1 Strain Gauges

Figure 5.1 shows the structure of a strain gauge together with a circuit with which the dependence of the sensor resistance on the strain can be simulated. The carrier foil is shown with the measuring grid etched out of a thin metal foil.

If a conductor is stretched, it becomes longer and thinner, which increases its resistance. The relative resistance change dR/R of the strain gauge is described in Eq. (5.1)

$$\frac{dR}{R} = k \cdot \frac{dl}{l} = k \cdot \varepsilon \tag{5.1}$$

with the nominal resistance R, the strain ε, the factor for strain sensitivity k and the relative change in length dl/l.

Data Sheet

Strain gage type LY11-6/120 from Hottinger/Baldwin GmbH [1] L: linear strip, LY: indication of a measuring grid made of *constantan* (60% Cu, 40% Ni). The factor for constantan is $k \approx 2$. Temperature coefficient: $TK_k \approx (115 +/- 10) \cdot 10^{-6}$ 1/K.

LY1: Measuring grid for strain measurements in one direction,
LY11: adapted to the thermal expansion of steel with $TC_{Stahl} = 10.8 \cdot 10^{-6}$ 1/K.
LY11-6: Strain gauge with the measuring grid length $a = 6$ mm, see Fig. 5.1.
LY11-6/120: means $R = 120 \, \Omega$.

Maximum bridge voltage for this strain gauge: $U_B = 8$ V.

© The Author(s), under exclusive license to Springer Fachmedien Wiesbaden GmbH, part of Springer Nature 2023
P. Baumann, *Selected Sensor Circuits*,
https://doi.org/10.1007/978-3-658-38212-4_5

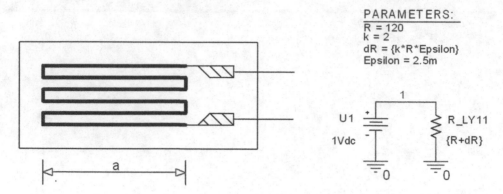

Fig. 5.1 Structure of a foil strain gauge and circuit for type LY11-6/120

Task: Characteristic Curve

To be shown is the relative change in resistance dR/R as a function of strain ε for the range $\varepsilon = -2.5$–10^{-3} to 2.5–10^{-3} according to [2].

Analysis

- PSpice, Edit Simulation Profile
- Simulation Settings – Fig. 5.1: Analysis
- Analysis type: DC Sweep
- Options: Primary Sweep
- Sweep variable: Global Parameter
- Parameter Name: Epsilon
- Sweep type: Linear
- Start value: −2.5 m
- End value: 2.5 m
- Increment: 10 u
- Apply: OK
- PSpice, run

At the reference temperature $T_0 = 23\,°C$, $k = 2$ and $dR/R = k{-}\varepsilon$. In the case of a strain, ε is positive, in the case of a compression, ε becomes negative.

The analysis result according to Fig. 5.2 shows that the relative resistance change dR/R increases linearly with the strain ε.

5.2   Structural Steel Bending Bar

Figure 5.3 shows a loaded bending bar with four constantan strain gauges glued on.

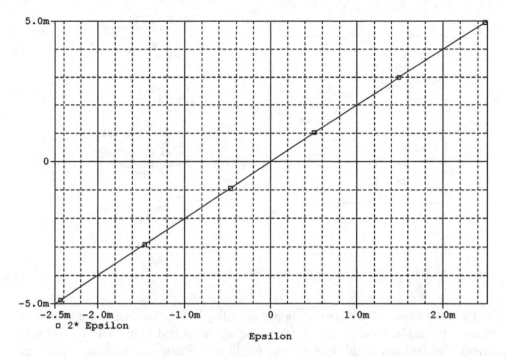

Fig. 5.2 Relative change in resistance as a function of strain

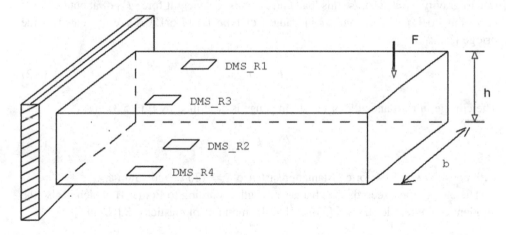

Fig. 5.3 Flexure bar with four strain gauges

    The strain gages located on the upper side of the bar are stretched and those located on the lower side are compressed. This causes the resistances R_1 and R_3 to increase by dR, while the resistances R_2 and R_4 decrease by dR. These resistance changes can be evaluated with a Wheatstone bridge.

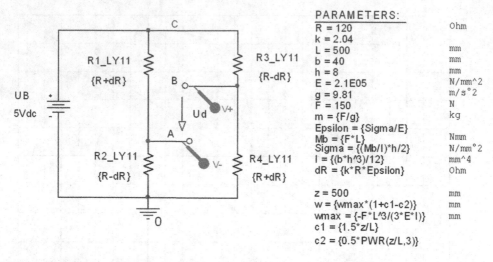

R = 120		Ohm
k = 2.04		
L = 500		mm
b = 40		mm
h = 8		mm
E = 2.1E05		N/mm^2
g = 9.81		m/s°2
F = 150		N
m = {F/g}		kg
Epsilon = {Sigma/E}		
Mb = {F*L}		Nmm
Sigma = {(Mb/I)*h/2}		N/mm°2
I = {(b*h^3)/12}		mm^4
dR = {k*R*Epsilon}		Ohm
z = 500		mm
w = {wmax*(1+c1-c2)}		mm
wmax = {-F*L^3/(3*E*I)}		mm
c1 = {1.5*z/L}		
c2 = {0.5*PWR(z/L,3)}		

Fig. 5.4 Measuring bridge for evaluating the bending load

The parameters of the measuring bridge according to Fig. 5.4 refer to the test with a bending bar made of mild steel [3], which was also evaluated in practice. The bar with length $L = 500$ mm, width $b = 40$ mm, height $h = 8$ mm and modulus of elasticity $E = 2.1–10^5$ N/mm^2 is loaded with a mass m of 0–15 kg. With the value of the acceleration due to gravity $g = 9.81$ m/s^2 this load corresponds to a weight force F_G of about 0–150 N according to Eq. (5.2). Four strain gauges of type LY11-6/120A are evaluated in the bridge [3].

$$F = m \cdot g \tag{5.2}$$

The change in resistance dR of the strain gauge is described by Eq. (5.3) via

$$dR = k \cdot R \cdot \varepsilon \tag{5.3}$$

with constant $k = 2.04$ for constantan, resistance $R = 120\ \Omega$ and strain ε.

The elongation ε records the change in length according to Eq. (5.4), which is also the quotient of the tensile stress σ [N/mm^2] to the modulus of elasticity E [N/mm^2].

$$\varepsilon = \frac{dl}{l} = \frac{\sigma}{E} \tag{5.4}$$

The tensile stress σ according to Eq. (5.5) is determined by the bending moment M_b, the area moment of inertia I of the bending bar, and the distance $h/2$ between the external and neutral phases of the material [4, 5].

$$\sigma = \frac{M_b}{I} \cdot \frac{h}{2}$$ (5.5)

The bending moment M_b is formed from the product of the force F and the bar length L, see Eq. (5.6).

$$M_b = F \cdot L$$ (5.6)

Equation (5.7) applies to the area moment of inertia I (in units of m^4) of the bending beam.

$$I = \frac{b \cdot h^3}{12}$$ (5.7)

The diagonal stress U_d of the full bridge according to Fig. 5.4 is obtained from Eq. (5.8) as follows

$$U_d = k \cdot \varepsilon \cdot U_B$$ (5.8)

Task: Bridge Voltage

With the circuit according to Fig. 5.4, the dependence of the bridge diagonal stress U_d on the weight force F (in the unit Newton) is to be represented. The length of the bending bar L is to be varied as 400 mm, 500 mm and 600 mm. The force shall be increased from 0 to 150 N.

Analysis

- PSpice, Edit Simulation Profile
- Simulation Settings – Fig. 5.4: Analysis
- Analysis type: DC Sweep
- Options: Primary Sweep
- Sweep variable: Global Parameter
- Parameter Name: F
- Sweep type: Linear
- Start value: 0
- End value: 150
- Increment: 10 m
- Options: Parametric Sweep
- Sweep variable: Global Parameter
- Parameter Name: L
- Sweep type: Value list: 400500600
- Edit diagram:

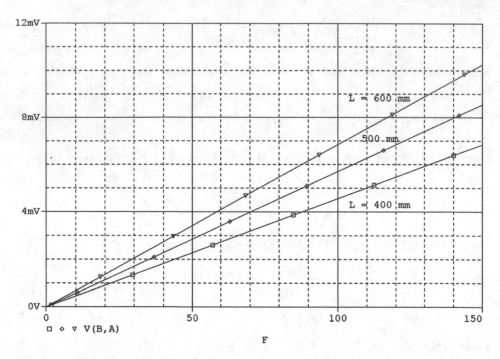

Fig. 5.5 Diagonal stress U_d as a function of weight force with length as parameter

- Trace, Add Trace
- Trace Expression: V(B,A)
- Apply: OK
- PSpice, run

The analysis result according to Fig. 5.5 shows for $L = 500$ mm the increase $dU_d/dF = 56.92$ μV/N.

Task: Mass and Weight Force

The relationship between the mass m and the weight force F for a change from 0 to 150 N is to be shown according to Eq. (5.2).

Analysis

- PSpice, Edit Simulation Profile
- Simulation Settings – Eq. (5.2): Analysis
- Analysis type: DC Sweep
- Options: Primary Sweep
- Sweep variable: Global Parameter
- Parameter Name: F

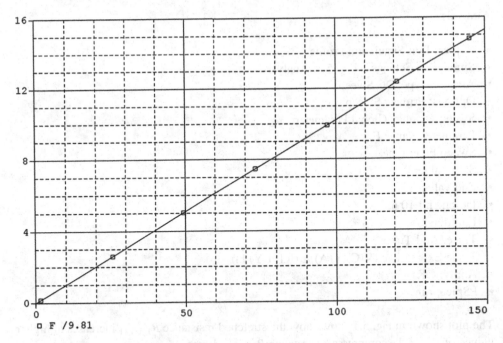

Fig. 5.6 Relationship between the mass in kilograms and the weight force in newtons

- Sweep type: Linear
- Start value: 0
- End value: 150
- Increment: 10 m
- Edit diagram:
- Trace, Add Trace
- Trace Expression: F/9.81
- Apply: OK
- PSpice, run

With the entry: F/9.81 the correlation between m and F emerges (Fig. 5.6).

Task: Resistance Changes

Using the bridge circuit shown in Fig. 5.4, the resistance changes dR with increasing weight force F_G can be analysed.

To be evaluated are: $R_{1\_LY11} = (U_C - U_A)/I(R_{1\_LY11})$ and $R_{2\_LY11} = U_A/IR2\_LY11$.

Analysis

- PSpice, Edit Simulation Profile
- Simulation Settings – Fig. 5.4: Analysis
- Analysis type: DC Sweep
- Options: Primary Sweep
- Sweep variable: Global Parameter
- Parameter Name: F
- Sweep type: Linear
- Start value: 0
- End value: 150
- Increment: 10 m
- Edit diagram:
- Trace, Add Trace
- Trace Expression: (V(C)-V(A))/I((R1_LY11))
- Apply: OK
- PSpice, run

The plot shown in Fig. 5.7 shows how the stretched resistance R_{1_LY11} increases at higher values of F and the compressed resistance R_{2_LY11} decreases.

Task: Elongation as a Function of the Weight Force
Using the circuit shown in Fig. 5.4, analyse how the strain ε depends on the force F.

Solution
From Eqs. (5.4), (5.5), (5.6) and (5.7), the strain follows from Eq. (5.9) with

$$\varepsilon = \frac{6 \cdot L \cdot F}{b \cdot E \cdot h^2} \tag{5.9}$$

One obtains $\varepsilon = 5.58\text{--}10^{-6}$ 1/N F.

Analysis

- PSpice, Edit Simulation Profile
- Simulation Settings – Fig. 5.4: Analysis
- Analysis type: DC Sweep
- Options: Primary Sweep
- Sweep variable: Global Parameter
- Parameter Name: F
- Sweep type: Linear
- Start value: 0

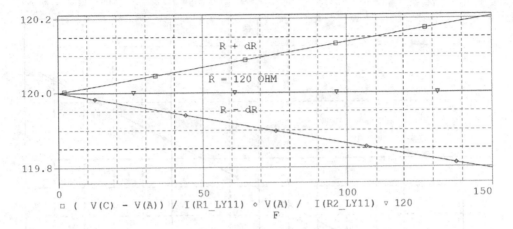

Fig. 5.7 Resistance increase with elongation and resistance decrease with compression

- End value: 150
- Increment: 10 m
- Edit diagram:
- Trace, Add Trace
- Trace Expression: F*5.58E-06
- Apply: OK
- PSpice, run

The dependence of the strain ε on the weight force F is shown in Fig. 5.8. Figure 5.8 shows that the strain increases linearly with the weight force.

Task: Bending Line
The bending line $w = f(x)$ of the structural steel strip with a left-sided load can be calculated using Eq. (5.10) [6]. The transformation for the right-sided loading according to Fig. 5.3 is done by changing the abscissa from z to $(-z + L)$, see Fig. 5.9.

$$w = \frac{-F \cdot L^3}{3 \cdot E \cdot I} \cdot \left[1 - \frac{3}{2} \cdot \left(\frac{z}{L}\right) + \frac{1}{2} \cdot \left(\frac{z}{L}\right)^3 \right] \tag{5.10}$$

For $z = L$, the maximum deflection $w_{\max}$ of the bending strip is given by Eq. (5.11) with

$$w_{max} = \frac{-F \cdot L^3}{3 \cdot E \cdot I} \tag{5.11}$$

One obtains $w_{\max} = 17.4386$ mm with $L = 500$ mm and $F = 150$ N.

The parameters of the circuit shown in Fig. 5.4 can be used to illustrate how the Deflection $w = f(x)$ for $z = 0$ to $z = L$ is performed.

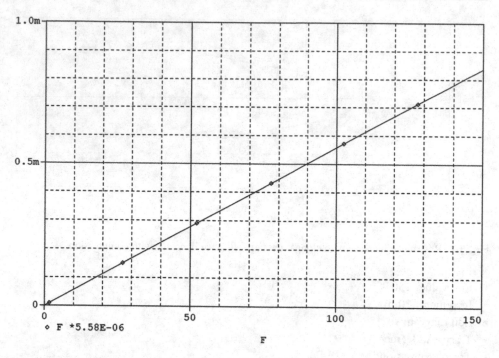

Fig. 5.8 Elongation as a function of the weight force in Newton

Analysis

- PSpice, Edit Simulation Profile
- Simulation Settings – Fig. 5.4: Analysis
- Analysis type: DC Sweep
- Options: Primary Sweep
- Sweep variable: Global Parameter
- Parameter Name: z
- Sweep type: Linear
- Start value: 0
- End value: 500
- Increment: 10 m
- Edit diagram:
- Plot, Axis Settings
- Axis variable: $-z + 500$
- Trace, Add Trace
- Trace Expression: $-17.439*(1-1.5*z/500 + 0.5\,PWR(z/500,3))$
- Apply: OK
- PSpice, run

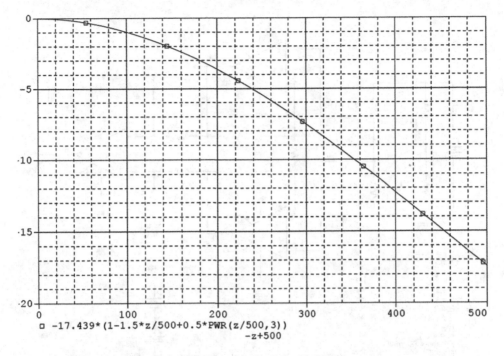

Fig. 5.9 Steel bar deflection in millimetres when loaded with 150 Newton

Figure 5.9 is obtained by entering the values according to Eq. (5.10) in Trace Expression. The maximum deflection calculated according to Eq. (5.11) is fulfilled.

The value of the diagonal voltage according to Fig. 5.4 can be amplified with an instrumentation amplifier, see Fig. 5.10.

The value of the diagonal voltage according to Fig. 5.4 can be amplified with an instrumentation amplifier, see Fig. 5.10.

For $R_6 = R_7$, $R_8 = R_9$ and $R_{10} = R_{11}$, the bridge voltage is amplified using Eq. (5.12) as follows:

$$U_Y = -\frac{R_{10}}{R_8} \cdot \left(1 + \frac{2 \cdot R_6}{R_5}\right) \cdot U_d \tag{5.12}$$

Using the values of the circuit of Fig. 5.10, the voltage gain is $v_u = 500$.

Task: Variation of the Force

For a variation of the weight force in the range $F = 0$ to 150 N, the diagonal voltage $U_d = U_B - U_A$ and the output voltage U_Y are to be represented.

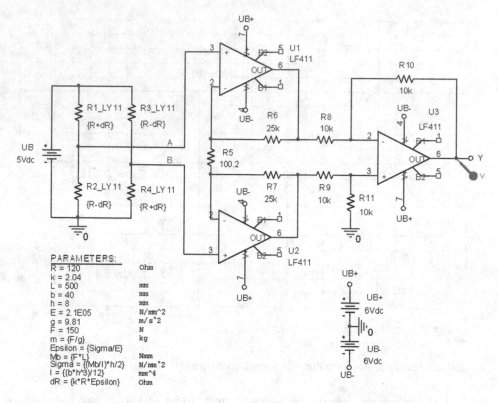

Fig. 5.10 Amplification of the bridge voltage with an instrumentation amplifier

Analysis

- PSpice, Edit Simulation Profile
- Simulation Settings – Fig. 5.4: Analysis
- Analysis type: DC Sweep
- Options: Primary Sweep
- Sweep variable: Global Parameter
- Parameter Name: F
- Sweep type: Linear
- Start value: 0
- End value: 150
- Increment: 5 m
- Edit diagram:
- Trace, Add Trace
- Trace Expression: V(B)-V(A) and V(Y)
- Apply: OK
- PSpice, run

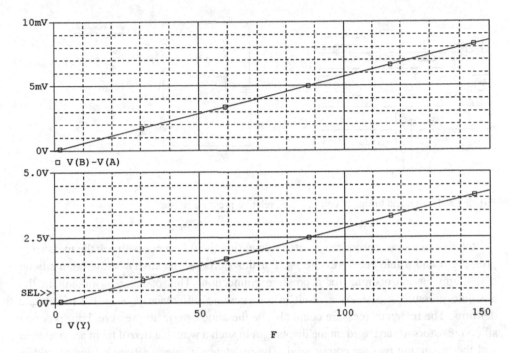

Fig. 5.11 Bridge and output voltage as a function of the weight force in Newton

With the analysis result of Fig. 5.11, the linear increase of bridge voltage and output voltage with weight force F is shown.

In the practically realized experiment, an AD 520 instrumentation amplifier, an ATMEGA 8 microcontroller, and a line display were used [3].

5.3   Piezoresistive P-Silicon Pressure Sensor

Narrow, thin p-doped silicon strips have a *k-factor of* $k \approx 100$, which is about 50 times higher than that of constantan gages. Under pressure, the resistivity ρ of silicon resistors changes greatly. The *k-factor* is obtained by Eq. (5.13) to:

$$k \approx \frac{d\rho/\rho}{dl/l} \qquad (5.13)$$

For p-silicon in the (111) direction and $p_{0p} \approx 2\text{--}10^{19}$ cm^{-3}, $k \approx 100$ at $T = 25$ °C [7].

The principle structure of a silicon pressure sensor with a bridge circuit is shown in Fig. 5.12, see [4, 5, 8, 9].

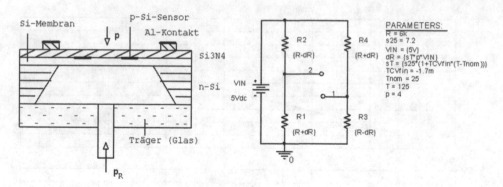

Fig. 5.12 Design and bridge circuit of a p-Si relative pressure sensor

Relative pressure sensors can be used to measure the difference between the measuring pressure p and a reference pressure p_R. For this purpose, the cavity below the n-silicon diaphragm has an opening for a pressure connection. This opening is missing in the absolute pressure sensor. As a result of the pressure difference $p_R - p$, the membrane deforms. The reference pressure could also be the atmospheric air pressure. Four p-doped silicon resistors are arranged on the diaphragm in such a way that two of them are stretched and the remaining two are compressed. The resistance changes dR are evaluated with a bridge, see Eq. (5.14).

Data Sheet
Piezoresistive silicon relative pressure sensor KPY 44-R from Infineon [10].
Operating temperature range $T_A = (-40 \ldots 120)$ °C, maximum input voltage: $V_{IN} = 12$ V.

Electrical characteristics at $T_A = 25$ °C, $V_{IN} = 5$ V: Sensitivity $s = 6$ (4 ... 9) Ω/(V-bar), Bridge resistance $R = 4 \ldots 8$ kΩ, output voltage: $V_{fin} = 120$ (80 ... 180) mV, Temperature coefficients $TC_R = 0.095\%$/K and $TC_{Vfin} = -0.17$ ($-0.19 \ldots -0.14$) %/K.

To obtain $V_{fin} = 120$ mV at $R = 6$ kΩ, $V_{IN} = 5$ V and $p = 4$ bar.
$s_{25} = 7.2$ Ω/(V-bar) is chosen. The change in resistance dR is calculated according to Eq. (5.14).

$$dR = s \cdot p \cdot V_{IN} \qquad (5.14)$$

The dependence of the sensitivity s on the temperature is described by Eq. (5.15) as follows

$$s = s_{25} \cdot \left(1 + TC_{Vfin} \cdot (T - T_{nom})\right) \tag{5.15}$$

with sensitivity $s = s_{25}$ at $T = T_{nom}$.

The output voltage (diagonal voltage of the bridge) is given by Eq. (5.16):

$$V_{fin} = V(2) - V(1) = \frac{dR}{R} \cdot V_{IN} \tag{5.16}$$

Task: Pressure Dependence of the Output Voltage

With the circuit for the full bridge according to Fig. 5.12, the pressure dependence of the output voltage V_{fin} for the range $p = 0$ to 4 bar can be analysed with:

(a) a variation of the input voltage with $V_{IN} = 5$ V and 12 V at $T = T_{nom} = 25\ ^\circ$C
(b) a variation of the temperature with $T = (-40, 25, 125)\ ^\circ$C at $V_{IN} = 5$ V.

Note: in SPICE, the character s cannot be used for sensitivity because it is assigned for the execution of an integration, so s_T is set instead of s.

Analysis for (a) with V_{IN} as Parameter

- PSpice, Edit Simulation Profile
- Simulation Settings – Fig. 5.12: Analysis
- Analysis type: DC Sweep
- Options: Primary Sweep
- Sweep variable: Global Parameter
- Parameter Name: p
- Sweep type: Linear
- Start value: 0
- End value: 4
- Increment: 1 m
- Options: Parametric Sweep
- Sweep variable: Global Parameter
- Parameter Name: VIN
- Sweep type: Value list: 5, 12
- Takeover: OK
- PSpice, run

(a) Parametric Sweep, Voltage Source, Parameter Name: V_{IN}, value list: 5, 12
(b) Parametric sweep, Global Parameter, Parameter Name: T, value list: −40, 25, 125.

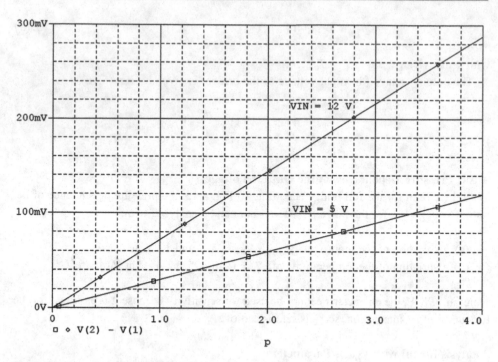

Fig. 5.13 Output voltage as a function of pressure

The analysis result according to Fig. 5.13 shows that the output voltage V_{fin} increases linearly with the pressure p and is proportional to the input supply voltage V_{IN}.

Note: 1 bar = 100 kPa = 10^5 N/m^2 = 1.02 kp/cm^2.

Analysis for (b) with T as Parameter

- PSpice, Edit Simulation Profile
- Simulation Settings – Fig. 5.12: Analysis
- Analysis type: DC Sweep
- Options: Primary Sweep
- Sweep variable: Global Parameter
- Parameter Name: p
- Sweep type: Linear
- Start value: 0
- End value: 4
- Increment: 1 m
- Options: Parametric Sweep
- Sweep variable: Global Parameter
- Parameter Name: T
- Sweep type: Value list: −40 25 125

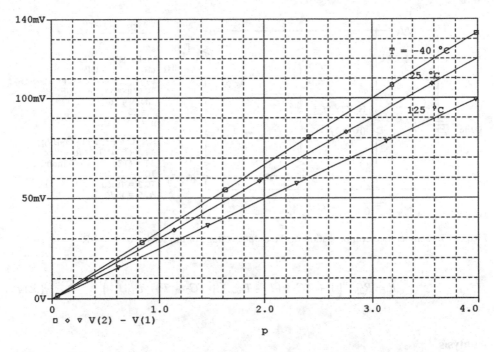

Fig. 5.14 Pressure dependence of the output voltage with temperature T as parameter

- Takeover: OK
- PSpice, run

The characteristic curves according to Fig. 5.14 show that the output voltage V_{fin} decreases with higher temperature. The (positive) temperature coefficient of the p-Si resistor R does not affect the output voltage V_{fin} in the full bridge with the four resistors because of their mutual compensation. However, the temperature dependence of the n-Si diaphragm should be noted [11].

In general, temperature compensation is required for the output voltage (as well as for the offset voltage U_O).

Task: Temperature Compensation
The decrease in output voltage with an increase in temperature must be compensated with a PTC resistor (temperature sensor). The pressure dependence of the output voltage V_{fin} for the pressure $p = 0$ to 4 bar is to be analyzed with $T = 25\,^\circ\text{C}$ and $125\,^\circ\text{C}$ as parameters, see Fig. 5.15.

The temperature dependence of the silicon PTC resistor sensor KTY11_5 with its positive temperature coefficients TC_1 and TC_2 is described by Eq. (5.17).

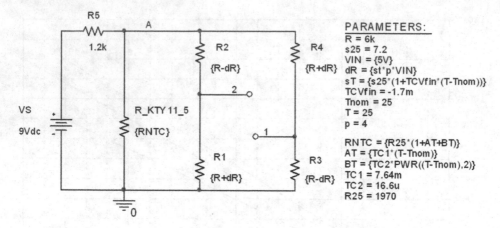

Fig. 5.15 Temperature compensation of the output voltage of the sensor KPY 44-R

$$R = R_{25} \cdot \left(1 + TC_1 \cdot (T - T_{nom}) + TC_2 \cdot (T - T_{nom})^2\right) \qquad (5.17)$$

Analysis

- PSpice, Edit Simulation Profile
- Simulation Settings – Fig. 5.15: Analysis
- Analysis type: DC Sweep
- Options: Primary Sweep
- Sweep variable: Global Parameter
- Parameter Name: p
- Sweep type: Linear
- Start value: 0
- End value: 4
- Increment: 1 m
- Options: Parametric Sweep
- Sweep variable: Global Parameter
- Parameter Name: T
- Sweep type: Value list: 25125
- Takeover: OK
- PSpice, run

Because the temperature dependence of the sensitivity s_T had to be made with the temperature T as a *global* parameter, the PTC resistor can also be described in this way

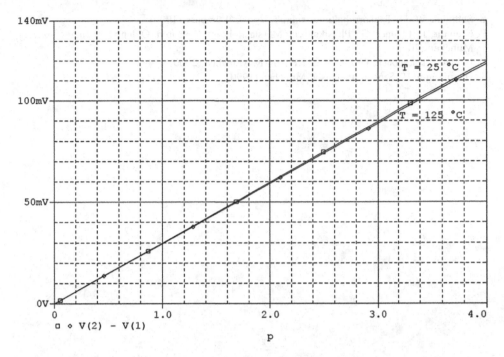

Fig. 5.16 Pressure dependence of the output voltage at two temperatures

as a temperature-dependent resistor. The temperature coefficients of the KTY11_5 are: $TC_1 = 7.64–10^{-3}$ 1/K and $TC_2 = 16.6–10^{-6}$ 1/K^2.

Analysis Result

With the selected series resistor $R_5 = 1.2$ kΩ, a signal voltage of $V_S = 9$ V in order to obtain the input voltage $V_{IN} = 5$ V at node A.

The temperature compensation according to Fig. 5.16 is successful, see Fig. 5.14 for comparison.

References

1. Hottinger Baldwin Messtechnik: Dehnungsmessstreifen. Katalog (2014)
2. Baumann, P.: Sensorschaltungen. Vieweg+Teubner, Wiesbaden (2010)
3. Brunßen, F., Kleen, J., Naumann, H., Kleesik, S.: DMS am Biegestab. Projektarbeit Hochschule, Bremen (2014)
4. Schnell, G.: Sensoren in der Automatisierungstechnik. Vieweg, Braunschweig (1991)
5. Hesse, S., Schnell, G.: Sensoren für die Prozess- und Fabrikautomation. Vieweg+Teubner, Wiesbaden (2009)
6. Hering, E., Modler, K.-H.: Grundwissen des Ingenieurs. fv. Leipzig. (2002)
7. Elbel, T.: Mikrosensorik. Vieweg+Teubner, Wiesbaden (1996)

8. Schmidt, W.-D.: Sensorschaltungstechnik. Vogel, Würzburg (1997)
9. Niebuhr, J., Lindner, G.: Physikalische Messtechnik mit Sensoren. Oldenbourg Industrieverlag, München (2011)
10. Infineon: Datenbuch. Drucksensor KPY 44-R, München (2000)
11. Hauptmann, P.: Sensoren. Hanser, München (1990)

Hall Sensor

6

6.1 Mode of Operation and Characteristic Curves

The Hall effect states: If a current-carrying semiconductor plate (GaAs, InAs) is vertically penetrated by a magnetic field, the charge carriers are deflected from their horizontal path, causing a Hall voltage U_2 to appear on the side surfaces, see Fig. 6.1. With an unloaded output with $I_2 = 0$, the open-circuit Hall voltage U_{20} appears according to Eq. (6.1):

$$U_{20} = K_{B0} \cdot B \cdot I_1 \tag{6.1}$$

with the open-circuit sensitivity K_{B0} in V/(A−T), the magnetic induction (flux density) B in T (Tesla) and the input control current I_1 in mA.

Note: 1 Tesla $= 1$ T $= 1$ Vs/m^2.

If the magnetic field does not act perpendicularly, but at an angle α, then Eq. (6.2) holds with:

$$U_{20} = K_{B0} \cdot B \cdot I_1 \cdot cos\alpha \tag{6.2}$$

Data Sheet

KSY10, Hall sensor, ion-implanted monocrystalline GaAs, Siemens [1].

$I_{1max} = 7$ mA, maximum control current, $I_{1N} = 5$ mA, nominal control current $K_{B0} = 200$, (170 ... 230) V/(A-T), no-load sensitivity $U_{20} = 85 ... 130$ mV at $I_1 = 5$ mA and $B = 0.1$ T, open circuit Hall voltage $R_{10}, R_{20} = 1$ kΩ, (900 ... 1200) Ω at $B = 0$ T, input and output resistance. Hall sensors are used for measuring magnetic fields as well as for position and motion detection [2–4].

© The Author(s), under exclusive license to Springer Fachmedien Wiesbaden GmbH, part of Springer Nature 2023
P. Baumann, *Selected Sensor Circuits*,
https://doi.org/10.1007/978-3-658-38212-4_6

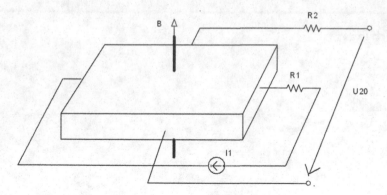

Fig. 6.1 Basic representation of the Hall sensor

Task: Dependence of the Hall Voltage on the Load Resistance

For the Hall sensor KSY10, the dependence of the Hall voltage U_2 on the load resistance R_L is to be shown with the circuits according to Fig. 6.2 for $R_L = 1\ \Omega$ to 1 MΩ and compared with the open-circuit Hall voltage U_{20}. The magnetic induction is to be varied with $B = 0.1$ T to 0.5 T in steps of 0.1 T.

Analysis

- PSpice, Edit Simulation Profile
- Simulation Settings – Fig. 6.2: Analysis
- Analysis type: DC Sweep
- Options: Primary Sweep
- Sweep variable: Global Parameter
- Parameter Name: RL
- Sweep type: Logarithmic, Decade
- Start value: 1
- End value: 1 Meg
- Points/Decade: 100
- Options: Parametric Sweep
- Sweep variable: Global Parameter
- Parameter Name: B
- Sweep type: Linear
- Start value: 0.1
- End value: 0.5
- Increment: 0.1
- Apply: OK
- PSpice, run

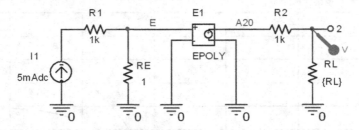

E^@REFDES %3 %4 VALUE={KB0*V(E)/RE*B}

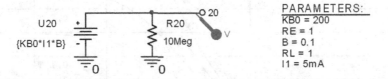

Fig. 6.2 Simulation circuits for the Hall sensor

The analysis result according to Fig. 6.3 shows that the Hall voltage U_2 changes to the open-circuit Hall voltage U_{20} at very large load resistances R_L. Note: in Fig. 6.2 (upper circuit) $U_{20} = U_{A20}$.

Task: Dependence of the Hall Voltage on the Magnetic Induction
Using the circuit shown in Fig. 6.2, the dependence of the open-circuit Hall voltage U_{20} on the magnetic induction for $B = 0$ to 0.5 T is to be analysed. For this purpose, the control current with the values $I_1 = 1$ mA, 3 mA, 5 mA and 7 mA must be used as parameters.

Analysis

- PSpice, Edit Simulation Profile
- Simulation Settings – Fig. 6.2: Analysis
- Analysis type: DC Sweep
- Options: Primary Sweep
- Sweep variable: Global Parameter
- Parameter Name: B
- Sweep type: Linear
- Start value: 0
- End value: 0.5
- Increment: 1 m
- Options: Parametric Sweep
- Sweep variable: Global Parameter
- Parameter Name: I1
- Sweep type: Linear

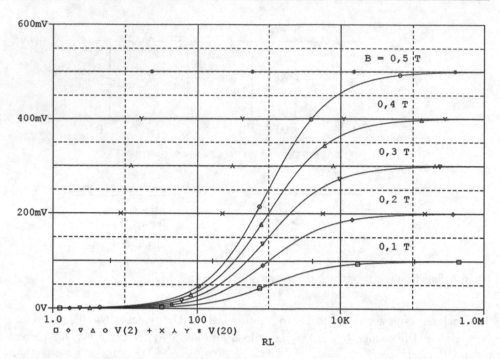

Fig. 6.3 Hall voltages as a function of the load resistance

- Start value: 1 mA
- End value: 7 mA
- Increment: 2 mA
- Apply: OK
- Pspice, run

The analysis result according to Fig. 6.4 states that the open-circuit Hall voltage U_{20} increases linearly with the induction B and the control current I_1 according to Eq. (6.1).

The data sheet specification for the open-circuit Hall voltage for magnetic induction $B = 0.1$ T and the nominal control current $I_{1N} = 5$ mA is fulfilled with $U_{20} = 100$ mV.

Task: Distance Measurement

Using the circuit shown in Fig. 6.5, the dependence of the open-circuit Hall voltage U_{20} on the distance a can be shown for the case where a permanent magnet with horizontal magnetisation moves away from a fixed Hall sensor. The magnetic induction would be $B = 0.35$ T. This simulation is based on the evaluations of the distance measurement with the sensor KSY10 given in [2].

The dependence of the open-circuit Hall voltage U_{20} on the distance a in units of millimeters can be captured by Eq. (6.3).

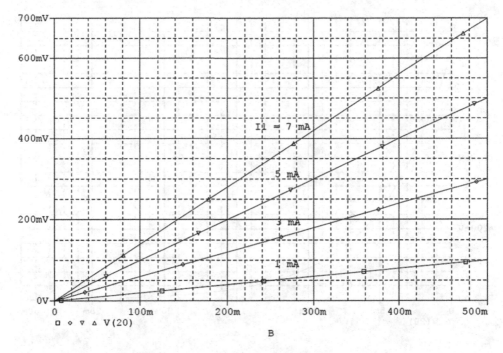

Fig. 6.4 Open-circuit Hall voltage as a function of induction with I_1 as parameter

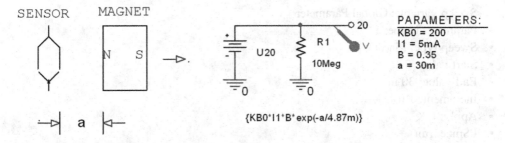

Fig. 6.5 Distance measurement with sensor KSY10

$$U_{20(a)} = K_{B0} \cdot I_1 \cdot B \cdot \exp\left(-a/4{,}87 \cdot 10^{-3}\right) \qquad (6.3)$$

Analysis

- PSpice, Edit Simulation Profile
- Simulation Settings – Fig. 6.5: Analysis
- Analysis type: DC Sweep
- Options: Primary Sweep

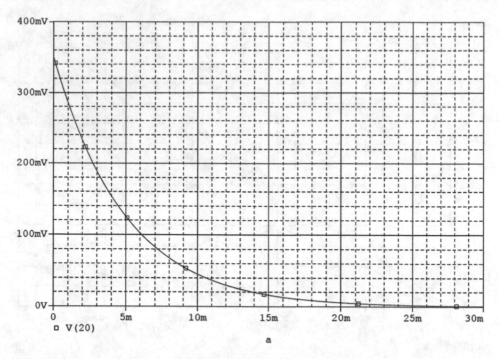

Fig. 6.6 Open-circuit Hall voltage as a function of the air gap distance in millimetres

- Sweep variable: Global Parameter
- Parameter Name: a
- Sweep type: Linear
- Start value: 0
- End value: 30 m
- Increment: 1 u
- Apply: OK
- PSpice, run

Note: For SPICE, the value 30 m corresponds to the value of 30 mm.

With the analysis result shown in Fig. 6.6, the representation given in [2] is replicated.

6.2   Hall Switch

The Hall switch shown in Fig. 6.7 consists of the Hall sensor with the parameters of type KSY 10, a non-feedback operational amplifier and a non-inverting Schmitt trigger, see also [3]. The Hall sensor supplies the open-circuit Hall voltage according to Eq. (6.1) to the input of the operational amplifier, which is constructed as a DC model with a voltage-controlled voltage source, a differential input resistor and an output resistor [5].

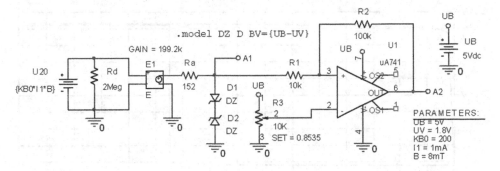

Fig. 6.7 Hall switch

These three model elements have the values of the operational amplifier μA 741:

- No-load differential gain (GAIN) vd0 = 106 dB or vd0 = 199.2–103
- Differential input resistance Rd. = 2–106 Ω = 2 MΩ
- Output resistance Ra = 152 Ω

The two Z-diodes serve to limit the output voltage to the level of the saturation voltages. In the absence of a magnetic field, i.e. when $B = 0$, the output voltage has the value $U_{A1} = 0$. In the ideal model, offset compensation is therefore unnecessary. If the offset voltage of the μA741 model (which is very low anyway) of $U_{OS} = -20\ \mu V$ were taken into account, then there would already be a deviation to $U_{A1} = 533\ \mu V$. As the magnetic induction increases, the Hall voltage increases so that the output A_1 of the first operational amplifier goes from LOW to HIGH. Depending on the reference voltage at the N input, which is set with the potentiometer R_3, the output A_2 of the non-inverting Schmitt trigger also switches from LOW to HIGH.

Task: Dependence of the Output Voltages on the Magnetic Field
The output voltages U_{A1} and U_{A2} shall be plotted as a function of magnetic induction (flux density) for the range $B = 0$ to 8 mT.

Analysis

- PSpice, Edit Simulation Profile
- Simulation Settings – Fig. 6.7: Analysis
- Analysis type: DC Sweep
- Options: Primary Sweep
- Sweep variable: Global Parameter
- Parameter Name: B
- Sweep type: Linear
- Start value: 0

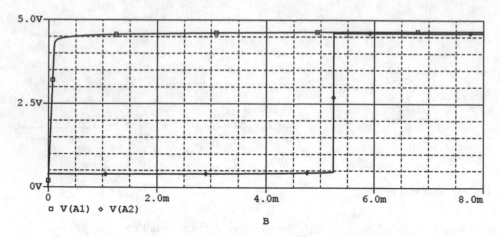

Fig. 6.8 Dependence of the output voltages on the magnetic induction

- End value: 8 mT
- Increment: 5 uT
- Apply: OK
- PSpice, run

The analysis result shown in Fig. 6.8 indicates that the Schmitt trigger is switched at $B \approx 5.2$ mT.

Task: Variation of the Control Current

For the circuit shown in Fig. 6.7, vary the control current with the flux density dependence of the output voltages with the values $I_1 = 1$ mA and 2 mA.

Analysis

- PSpice, Edit Simulation Profile
- Simulation Settings – Fig. 6.7: Analysis
- Analysis type: DC Sweep
- Options: Primary Sweep
- Sweep variable: Global Parameter
- Parameter Name: B
- Sweep type: Linear
- Start value: 0
- End value: 8 mT
- Increment: 5 uT
- Options: Parametric Sweep
- Sweep variable: Global Parameter

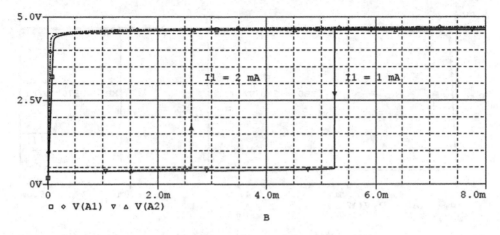

Fig. 6.9 Course of the output voltages with a variation of the control current

- Parameter Name: I1
- Sweep type: Value list: 1 mA, 2 mA
- Apply: OK
- PSpice, run

Because the Hall voltage increases with a higher control current according to Eq. (6.1), the output voltage U_{A2} in this case is already switched from LOW to HIGH at a smaller value of the magnetic induction B, see Fig. 6.9.

Task: Variation of the Idle Sensitivity

The circuit shown in Fig. 6.7 is to be considered. The dependence of the two output voltages on the magnetic induction is to be simulated with a variation of the no-load sensitivity in the values $K_{B0} = 170$ V/(A−T) and 230 V(A−T). This scattering range of K_{B0} corresponds to the specifications of the data sheet for the Hall sensor KSY 10 according to [1].

Analysis

- PSpice, Edit Simulation Profile
- Simulation Settings – Fig. 6.7: Analysis
- Analysis type: DC Sweep
- Options: Primary Sweep
- Sweep variable: Global Parameter
- Parameter Name: B
- Sweep type: Linear
- Start value: 0

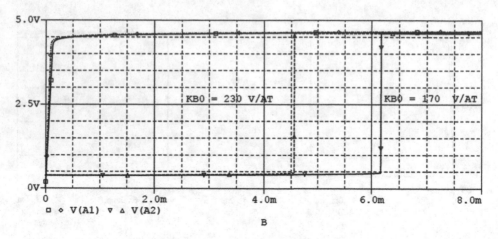

Fig. 6.10 Output voltages with a variation of the no-load sensitivity

- End value: 8 mT
- Increment: 5 uT
- Options: Parametric Sweep
- Sweep variable: Global Parameter
- Name: KB0
- Sweep type: Value list: 170, 230
- Apply: OK
- PSpice, run

According to Eq. (6.1), the larger value of the open-circuit sensitivity results in the higher Hall voltage, which means that the switching of the output voltages occurs at a lower magnetic induction, see Fig. 6.10.

6.3   Switching Hysteresis

In the circuit shown in Fig. 6.11, the Hall voltage $U_{20} = K_{B0} - B - I_1$ according to Eq. (6.1) is at the input of a non-inverting Schmitt trigger (with positive and negative operating voltage).

The switching points of this input voltage are captured by Eqs. (6.4) and (6.5).

$$U_{E1} = -U_{S-} \cdot \frac{R_1}{R_2} \tag{6.4}$$

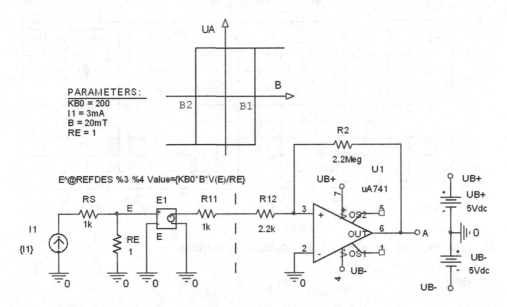

Fig. 6.11 Control of a Schmitt trigger by a Hall sensor

$$U_{E2} = -U_{S+} \cdot \frac{R_1}{R_2} \tag{6.5}$$

The hysteresis width follows from Eq. (6.6) with:

$$U_{HB} = (U_{S+} - U_{S-}) \cdot \frac{R_1}{R_2} \tag{6.6}$$

When the voltage supply of the operational amplifier is $U_B = \pm 5$ V, the positive saturation voltage reaches $U_{+S} = 4.57$ V and the negative saturation voltage arrives at $U_{-S} = -4.57$ V. Using Eqs. (6.4) and (6.5) and $R_1 = R_{11} + R_{12}$, the switching thresholds of the input voltage corresponding to the open-circuit reverberation voltage are obtained to be $U_{E1} = 6.65$ mV and $U_{E2} = -6.65$ mV. From Eq. (6.6), the hysteresis width is obtained as $U_{HB} = 13.3$ mV. The switching thresholds for magnetic induction can be obtained from the conversion of Eq. (6.1). Using the parameter values from Fig. 6.1, the switching points are obtained with the values $B_1 = 11.1$ mT and $B_2 = -11.1$ mT. The hysteresis width follows with $B_{HB} = 22.2$ mT.

Task: Dependence of the Output Voltage on the Magnetic Induction

For the circuit shown in Fig. 6.11, the dependence of the output voltage on the magnetic induction B is to be analysed. B is first to be increased from -20 mT to 20 mT and then reduced to -20 mT in a subsequent analysis starting at 20 mT.

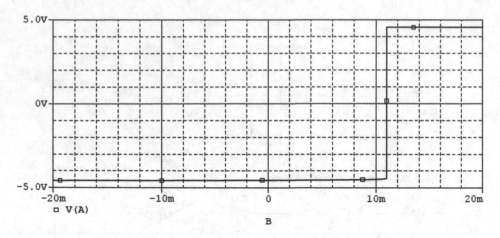

Fig. 6.12 Switching the output voltage with increasing induction

Analysis

- PSpice, Edit Simulation Profile
- Simulation Settings – Fig. 6.11: Analysis
- Analysis type: DC Sweep
- Options: Primary Sweep
- Sweep variable: Global Parameter
- Parameter Name: B
- Sweep type: Linear
- Start value: −20 m
- End value: 20 mT
- Increment: 5 uT
- Apply: OK
- PSpice, run

For an increasing magnetic induction B the analysis result is the diagram according to Fig. 6.12 with the switching threshold $B_1 = 11.1$ mT.

In the opposite direction, i.e. with Start Value $= 20$ mT and End Value $= -20$ mT, the switching threshold is $B_2 = -10.6$ mT, see Fig. 6.13.

The hysteresis width thus follows to $B_{HB} = 21.7$ mT. Approximately, this confirms the previous calculation.

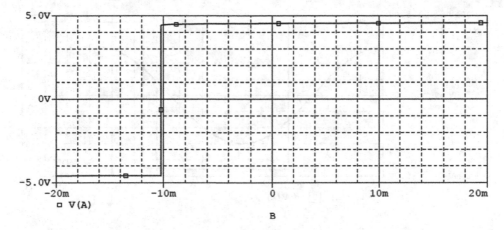

Fig. 6.13 Switching the output voltage with decreasing induction

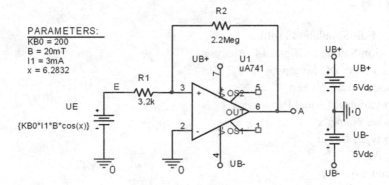

Fig. 6.14 Controlling the Schmitt trigger with a cosine Hall voltage

6.4 Hall Voltage with Cosine Curve

In the circuit shown in Fig. 6.14, a Hall voltage is applied to the Schmitt trigger which has a cosine dependence in order to be able to pass through the hysteresis characteristic.

Task: Hysteresis for Cosine Hall Voltage at Schmitt Trigger

For the circuit shown in Fig. 6.14, the dependence of the output voltage U_A on the input voltage U_E as well as on the magnetic induction B for one period of the cosine function is to be shown.

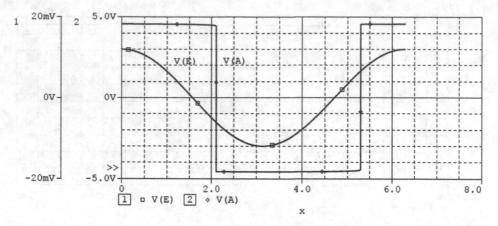

Fig. 6.15 Input and output voltage as a function of the x-coordinate

Analysis

- PSpice, Edit Simulation Profile
- Simulation Settings – Fig. 6.14: Analysis
- Analysis type: DC Sweep
- Options: Primary Sweep
- Sweep variable: Global Parameter
- Parameter Name: x
- Sweep type: Linear
- Start value: 0
- End value: 6.2832
- Increment: 1 m
- Apply: OK
- PSpice, run

The analysis result according to Fig. 6.15 shows the cosine input voltage and the resulting square-wave output voltage.

In the next step, the abscissa variable is modified via Plot, Axis Settings, Axis variable from x to $U_E = V(E)$. The switching thresholds from Fig. 6.16 are at $U_{E1} = 6.1$ mV and $U_{E2} = -6.1$ mV, see also paragraph 6.3.

Finally, the hysteresis curve is transformed from $U_A = f(U_E)$ to $U_A = f(B)$ via $B = V(E)/(B*I_1)$ according to Eq. (6.1).

Figure 6.17 shows the switching thresholds $B_1 = 11$ mT and $B_2 = -10.2$ mT. These values largely agree with those from Sect. 6.3.

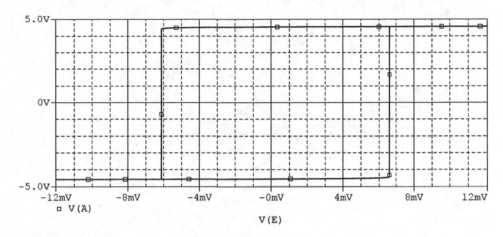

Fig. 6.16 Output voltage as a function of input voltage

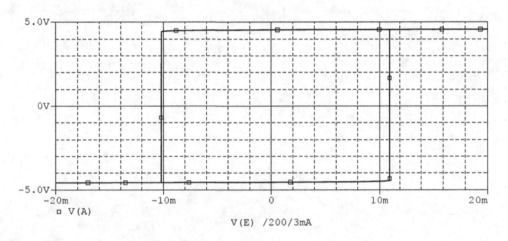

Fig. 6.17 Output voltage as a function of magnetic induction in milli-Tesla

References

1. Siemens: Datenblatt des Hallsensors KSY 10. München (1999)
2. Schnell, G.: Sensoren in der Automatisierungstechnik. Vieweg, Braunschweig (1991)
3. Böhmer, E., Ehrhardt, D., Oberschelp, W.: Elemente der angewandten Elektronik. Vieweg +Teubner, Wiesbaden (2010)
4. Baumann, P.: Sensorschaltungen. Vieweg+Teubner, Wiesbaden (2010)
5. Baumann, P.: Parameterextraktion bei Halbleiterbauelementen. Springer-Vieweg, Wiesbaden (2012)

Reed Components

7

To activate the reed contacts in the reed relay, a magnetic field is generated by the current in the relay coil. The switching operations are faster with these relays than with standard relays. Reed sensors are operated as reliable switches via a permanent magnet and serve as proximity switches, among other things. Reed components are widely used in measurement technology, automation and safety technology as well as in telecommunications.

7.1   Mode of Action

Reed components contain two contact tongues made of a soft magnetic alloy, which are placed in a glass tube filled with inert gas and surrounded by a coil, see Fig. 7.1. If an electric current flows through the coil or if a permanent magnet approaches the reed sensor, magnetic poles of different polarity are formed at the contact tongues due to the resulting magnetic field. For a sensor type with normally open contact (NO type), the electrical contact closes as soon as the magnetic force becomes greater than the restoring force of the contact tongues.

Data Sheet
DIL-SIL reed relay, TYPE 3570-1210 1 NO contact, type A, S.T.G. Germany GmbH [1].
    Rated voltage: 12 V, response voltage: max. 9 V, drop-out voltage: min. 1 V, coil resistance: 1 kΩ +/− 10%, max. switching capacity: 10 W/VA, max. switching current: 0.5 A, contact resistance: <150 mΩ, response time incl. bounce time max. 0.5 ms, drop-out time with diode: 0.5 ms. Since high self-induction voltages occur when switching off inductive loads, circuits are required to protect electronic components or to prevent damage

Fig. 7.1 Basic structure of the
reed component

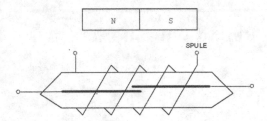

to the contacts. For DC circuits, free-wheeling diodes or varistors are used for this purpose,
while RC elements or varistors are used for AC circuits [2].

Task: Shutdown Behaviour

The relay coil bridged with a free-wheeling diode as shown in Fig. 7.2 should, after the time
$t = 0.1$ ms can be switched off. The course of the voltage across the node K of the circuit
breaker and the current through the inductance L_1, which is specified as constant, are to be
analysed.

Analysis

- PSpice, Edit Simulation Profile
- Simulation Settings – Fig. 7.2: Analysis
- Analysis type: Time Domain (Transient)
- Run to time: 700 us
- Start saving data after: 0 us
- Maximum step size: 10 us
- Apply: OK
- PSpice, run

The input voltage $U(E) = 12$ V appears in the analysis result according to Fig. 7.3. The
induced voltage increase which occurs briefly after switching off corresponds to the gate
voltage $U_{F0} = 0.67$ V of the freewheeling diode D1N 4007.

In the period $t = 0$ to 100 μs the switch is still closed and the current is $I = U_E/$
$R_1 = 12$ mA. When the contacts are opened, the coil current drops exponentially with the
time constant $\tau = L_1/R_1 = 100$ μs.

For more precise considerations, contact capacitances must be included. Without the
diode D_1 a voltage peak of 4.2 kV would occur for a short time!

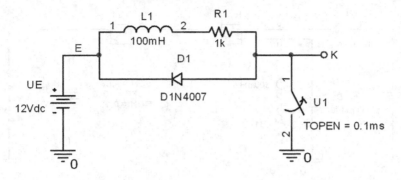

Fig. 7.2 Switching off an inductive load

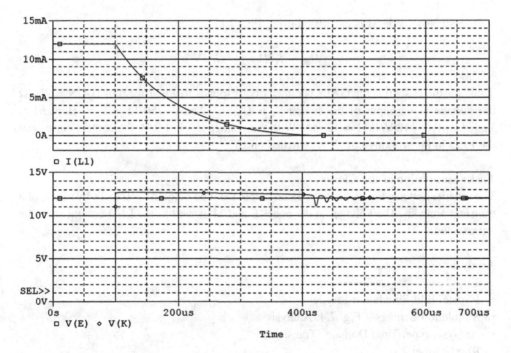

Fig. 7.3 Switch-off processes for coil current and contact voltage

7.2   Reed Relay as Normally Open Contact

In the circuit shown in Fig. 7.4, the control circuit consists of the on switch U_1, the coil with L_1 and R_1, and the free-wheeling diode D_1. The closing of the contacts is simulated by the interaction of the current-controlled voltage source H_1 (from the Analog library) with the voltage-controlled switch S (from the ANALOG library).

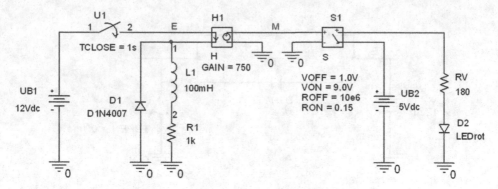

Fig. 7.4 Relay circuit with control circuit and operating circuit

The stress at node M is calculated using Eq. (7.1).

$$U_M = I(H1 : 1) \cdot GAIN \tag{7.1}$$

Thus, for the H source, the GAIN parameter corresponds to a resistor. For GAIN = 750 Ω the voltage at the input of the switch S_1 is $U_M = 12$ mA−750 $\Omega = 9$ V. Here I(H1: 1) = $U_{B1}/R_1 = 12$ V/1k$\Omega = 12$ mA. With the voltage level of 9 V, the contact closes and the LED in the working circuit is switched on.

Task: Simulation of the Closer
The circuit shown in Fig. 7.4 is to be examined in the time range $\Delta t = 0$ to 2 s by means of a transient analysis. The voltages at the nodes E and M as well as the LED current are to be represented.

Analysis

- PSpice, Edit Simulation Profile
- Simulation Settings – Fig. 7.4: Analysis
- Analysis type: Time Domain (Transient)
- Run to time: 2 s
- Start saving data after: 0 s
- Maximum step size: 10 ms
- Apply: OK
- PSpice, run

Evaluating the analysis, it can be seen that the timer U_1 switches on the voltage at node E with 12 V and at node M with 9 V after 1 s has elapsed (Fig. 7.5).

The LED current reaches $I(D_2) = (U_{B2} - U(D_2))/R_V = (5$ V − 1.63 V)/ 180 $\Omega = 18.72$ mA.

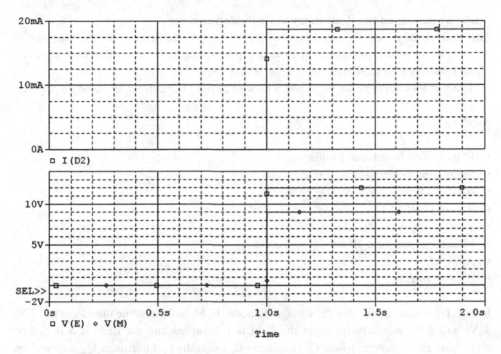

Fig. 7.5 Switching on the LED current and voltages at nodes E and M

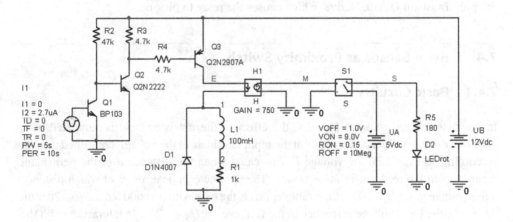

Fig. 7.6 Triggering the reed relay in case of a light interruption

7.3   Light Barrier with Reed Relay

The circuit shown in Fig. 7.6 causes the reed relay to switch on the LED when the light beam directed at the phototransistor is interrupted.

Task: Switching on the Reed Relay via a Phototransistor

The analysis of the circuit according to Fig. 7.6 shall be carried out for the time period $\Delta t = 0$ to 15 s. The photocurrent $I_2 = 2.7$ µA corresponds to an illumination of the phototransistor with the illuminance $E_v = 1000$ lx, see Sect. 3.7.

The voltages at the nodes E, M and S as well as the LED current are to be displayed.

Analysis

- PSpice, Edit Simulation Profile
- Simulation Settings – Fig. 7.6: Analysis
- Analysis type: Time Domain (Transient)
- Run to time: 15 s
- Start saving data after: 0 s
- Maximum step size: 10 ms
- Apply: OK
- PSpice, run

In Fig. 7.7, it can be seen that the voltages at nodes E, M, and S assume the values of 12 V, 9 V, and 5 V, respectively, when the light is interrupted and the LED is active. The non-illuminated phototransistor Q_1 is turned off, while the npn transistor Q_2 is turned on with the sufficiently high collector-emitter voltage of the phototransistor. This also makes the pnp transistor Q_3 conductive, which causes the relay to pick up.

7.4   Reed Sensor as Proximity Switch

7.4.1   Basic Circuitry

In the diagram according to Fig. 7.8, the effect of the magnetic field is simulated by a variable, distance-dependent voltage at the input terminals of the voltage-controlled switch. According to Eq. (7.2), the voltage U_S increases when the distance a of the permanent magnet from the reed sensor is decreased. The parameter n determines at which distance a the voltage value $U_S = 0$. In the example, this is the case with $n = 600$ V/m at $a = 20$ mm. At $a = 0$, the input voltage is reached at the full level of $U_S = 12$ V. In interaction with the switch characteristics, the contact closes at a certain distance, which switches on the LED.

$$U_S = U - n \cdot a \tag{7.2}$$

Task: Approach of the Magnet to the Reed Sensor

In Fig. 7.8 the distance a of the permanent magnet from the sensor is to be reduced from twenty to zero millimetres. Via the parameter h a variation with $V_{OFF} = 1$ V or 5 V is to be

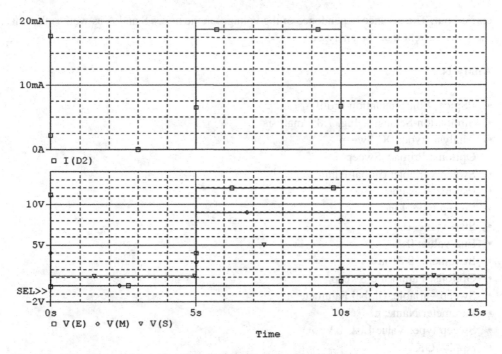

Fig. 7.7 Switching on three voltages and the LED during light interruption

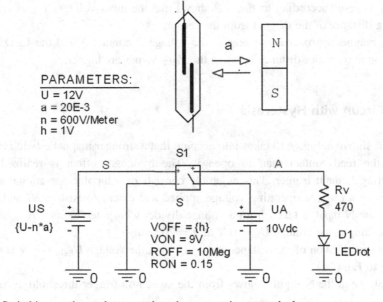

Fig. 7.8 Switching on the reed sensor when the magnet is approached

carried out. The distance dependence of the voltages at the nodes S and A as well as that of the LED current must be analysed.

Analysis

- PSpice, Edit Simulation Profile
- Simulation Settings – Fig. 7.8: Analysis
- Analysis type: DC Sweep
- Options: Primary Sweep
- Sweep variable: Global Parameter
- Parameter Name: a
- Sweep type: Linear
- Start value: 0.02
- End value: 0
- Increment: 0.01 m
- Options: Parametric Sweep
- Sweep variable: Global Parameter
- Parameter Name: h
- Sweep type: Value List. 1 V, 5 V
- Apply, OK
- PSpice, run

The analysis result according to Fig. 7.9 shows that the input voltages U_S increase with decreasing distance of the magnet from the sensor.

As the magnet approaches the sensor, the voltage at contact A and the LED are only switched on at a smaller distance a, when the V_{OFF} values are higher.

7.4.2   Circuit with Hysteresis

The circuit shown in Fig. 7.10 takes into account that a stronger magnetic field is required to close the reed contact than to open it. The hysteresis effect is realized with a non-inverting Schmitt trigger. This assembly consists of a bipolar operational amplifier LM 324 with a positive operating voltage applied and external resistors R_1 and R_2. The voltage at the N input is taken from a voltage divider with potentiometer R_{Pot}. As in the previous example, the voltage $U_E = 0$ V at $a = 20$ mm.

For the specification of the distance $a_{ein} = 10$ mm, the voltage $U_{Eein} = 6$ V is obtained according to Eq. (7.2).

The voltage at the N input follows from the switch-on trigger threshold according to Eq. (7.3).

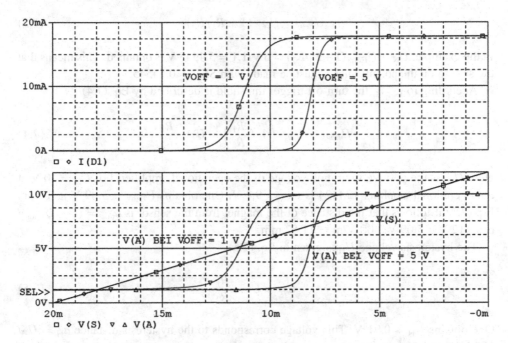

Fig. 7.9 Switch-on processes with voltage V_{OFF} as parameter

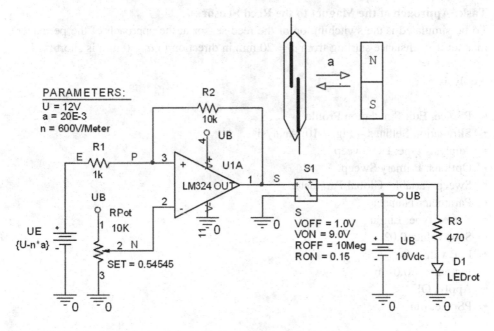

Fig. 7.10 Reed relay with simulation of hysteresis via a Schmitt trigger

$$U_{Eein} = U_N \cdot (1 + R_1/R_2) \tag{7.3}$$

In the example, $U_N = U_{Eein}/(1 + R_1/R_2) = 6 \text{ V}/1.1 = 5.4545$ V is obtained. This means that the setting on the potentiometer must be made with SET $= 0.54545$.

According to [3, 4], the turn-off trigger threshold is described by Eq. (7.4).

$$U_{Eaus} = U_N \cdot \left(1 + \frac{R_1}{R_2}\right) - U_{S+} \cdot \frac{R_1}{R_2} \tag{7.4}$$

Here U_{S+} is the positive saturation voltage of the operational amplifier. For $U_B = 10$ V at the operational amplifier LM324 is $U_{S+} = 9.1$ V. One obtains $U_{Eaus} = 5.09$ V and from Eq. (7.2) it follows that the distance of the magnet from the sensor is $a_{aus} = (U - U_{Eaus})/$ n $= (12 \text{ V} - 5.09 \text{ V})/600 \text{ V/m} = 11.52$ mm.

The hysteresis voltage U_H is given in Eq. (7.5).

$$U_H = U_{S+} \cdot \frac{R_1}{R_2} \tag{7.5}$$

One obtains $U_H = 0.91$ V. This voltage corresponds to the hysteresis distance $a_H = U_H/$ 600 V/m ≈ 1.52 mm.

Task: Approach of the Magnet to the Reed Sensor

To be simulated is the switching on of the reed sensor at the approach of the permanent magnet if its distance starting from $a = 20$ mm in direction to $a = 0$ mm is shortened.

Analysis

- PSpice, Edit Simulation Profile
- Simulation Settings – Fig. 7.10: Analysis
- Analysis type: DC Sweep
- Options: Primary Sweep
- Sweep variable: Global Parameter
- Parameter Name: a
- Sweep type: Linear
- Start value: 0.02
- End value: 0
- Increment: 0.01 m
- Apply, OK
- PSpice, run

The analysis shows that the reed contact is still open in the section $\Delta a = 20$ to 10 mm. Over the entire travel distance, the voltage at the N input of the operational amplifier V

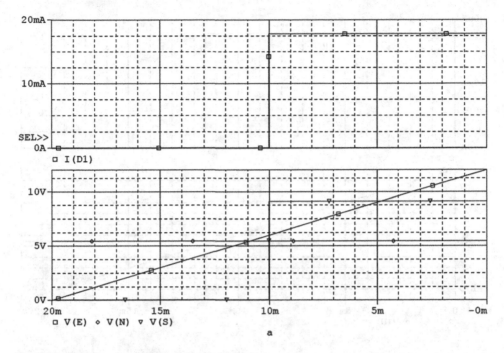

Fig. 7.11 Switch-on procedure of the reed relay

(N) = 5.4545 V. At a distance $a = 10$ mm, V(E) = V(S) = 6 V. At $a \leqq 10$ mm the contact closes and the LED lights up (Fig. 7.11).

Task: Removing the Magnet from the Reed Sensor
The switch-off process is to be simulated for the case that the magnet moves away from the reed sensor starting from the distance $a = 0$ in the direction of $a = 20$ mm.

Analysis

- PSpice, Edit Simulation Profile
- Simulation Settings – Fig. 7.10: Analysis
- Analysis type: DC Sweep
- Options: Primary Sweep
- Sweep variable: Global Parameter
- Parameter Name: a
- Sweep type: Linear
- Start value: 0
- End value: 20 m
- Increment: 0.01 m
- Apply, OK
- PSpice, run

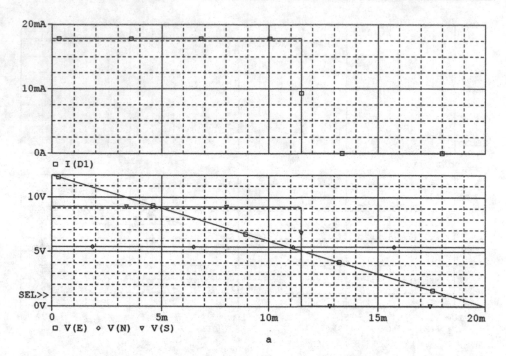

Fig. 7.12 Switching off the reed relay

From the analysis, it can be seen that the reed contact remains closed according to Eq. (7.4) over the now longer travel distance $\Delta a = 0$ to 11.52 mm (Fig. 7.12).

7.4.3   Representation of the Hysteresis Loop

To represent the hysteresis behaviour, a PWL source U_E is connected to the input in the circuit shown in Fig. 7.13.

At time $t = 0$ s $U_E = 12$ V and has dropped to $U_E = 0$ V at $t = 10$ s to reach $U_E = 12$ V at $t = 20$ s again.

Task: Hysteresis
For the time interval $\Delta t = 0$ to 20 s, the voltages at nodes E and S and the LED current are to be plotted. Finally, the time axis Time is to be converted to the distance a with the transformation from Eq. (7.2) according to $a = (U - V(E))/n$.

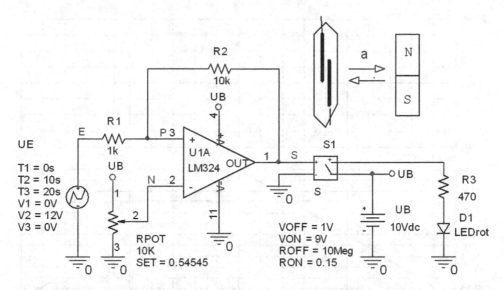

Fig. 7.13 Circuit for simulating the hysteresis loop

Analysis

- PSpice, Edit Simulation Profile
- Simulation Settings – Fig. 7.13: Analysis
- Analysis type: Time Domain (Transient)
- Options: General Settings
- Run to time: 20 s
- Start saving data after: 0 s
- Maximum step size: 1 ms
- Plot, Axis Settings, Axis variable: (12 V − V(E))/600 V
- Apply: OK
- PSpice, run

Figure 7.14 shows that the LED lights up in the time interval between the two intersections of the voltages U_E and U_S, see also Sect. 7.4.2. The LED is therefore active in the time interval in which the input voltage goes from $U_{Eein} = 6$ V to $U_{Eaus} = 5.09$ V. The LED lights up in the time interval between the two intersections of the voltages U and U, see also Sect. 7.4.2.

After converting the abscissa from time to distance a, it can be seen in Fig. 7.15 that the reed switch turns on at $a = 10$ mm, but does not turn off until the magnet has moved to distance $a − 11.52$ mm from the sensor.

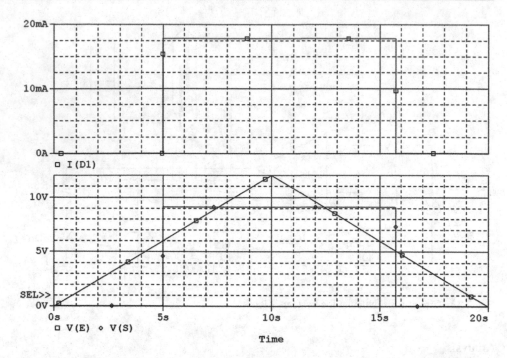

Fig. 7.14 Time dependence of voltages and LED current

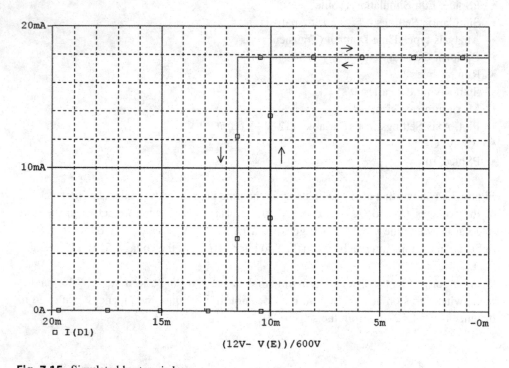

Fig. 7.15 Simulated hysteresis loop

References

1. S.T.G. Germany GmbH: DIL-SIL-REEDRELAIS, Datenblatt (2015)
2. Schiessle, E.: Industriesensorik. Vogel, Würzburg (2010)
3. Weddigen, C.H., Jüngst, W.: Elektronik. Springer Verlag, Berlin (1993)
4. Böhmer, E., Ehrhardt, D., Oberschelp, W.: Elemente der angewandten Elektronik. Vieweg +Teubner, Wiesbaden (2012)

Piezoelectric Buzzer

8

Piezoelectric buzzers are based on the reciprocal piezoelectric effect. Accordingly, if an alternating voltage is applied to an arrangement in which a PZT ceramic disc is connected to a brass disc, the ceramic transfers the vibrations generated by compression and expansion to the metal diaphragm and generates a buzzing tone in a narrow frequency range. Buzzers with two electrodes (connected to ceramic and metal ground) require external control via a square-wave or sine-wave generator (external drive). Buzzers with three electrodes use their third electrode for feedback and thus for self drive. In general, both types of transducers can be used as acoustic signal generators, for example in sensor circuits.

8.1 Buzzer for External Control

Buzzers with external control have two electrodes and are offered for different frequency ranges. Two buzzers for the frequencies of 2.9 and 4 kHz are considered first, the characteristic values of which are compiled from the data sheet in Table 8.1.

Figure 8.1 shows a sketch of the buzzer dimensions.

8.1.1 PSPICE Models of Buzzers with External Control

The procedures for determining the elements of the PSPICE models of piezoelectric buzzers with one or two C-L-R branches are described below.

P. Baumann, *Selected Sensor Circuits*,
https://doi.org/10.1007/978-3-658-38212-4_8

Table 8.1 Characteristics of piezoelectric buzzers from EKULIT [1]

Parameter	Unit	EPZ-27MS44W	EPZ-35MS29W
Frequency	kHz	4.4	2.9
Impedance	Ω	200	200
Capacity	nF	21	26
Diameter D	mm	27	35
Diameter d	mm	20	25
Thickness T	mm	0.53	0.56
Thickness t	mm	0.28	0.30

Fig. 8.1 Illustration of the slices of the buzzer

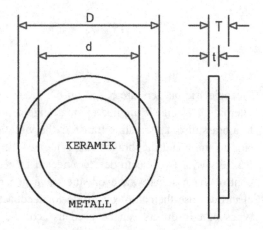

Buzzer EPZ-27 MS44

Figure 8.2 shows a circuit in which the buzzer is connected via the series resistor R_V to the signal generator with the rms voltage $U_E = 1$ V. The frequency-dependent output voltage U_A can be measured with an oscilloscope.

In the simplest case, the SPICE model of a piezoelectric transducer contains the series elements C_1, L_1 and R_1 as well as the parallel capacitance C_0. For such a model, the output voltage of the circuit shown in Fig. 8.2 has a minimum at frequency f_{s1} and a maximum at frequency f_{p1}.

As can be seen from the measurements, the buzzer of type EPZ-27MS44 exhibits such behaviour. Table 8.2 lists the resonant frequencies measured on a freely suspended specimen of this type, together with the associated amplitudes and the phase angle occurring during series resonance. In addition, the phase angle measured at $f = 1$ kHz is given as the sum of the resonant circuit capacitances C_0 and C_1.

The series capacitance C_1 and the parallel capacitance C_0 can be determined by Eqs. (8.1) and (8.2), respectively. On the one hand, Eq. (8.1) describes the quotient of the capacities with

Fig. 8.2 Circuit for measuring
the AC voltage across the buzzer

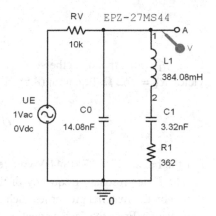

Table 8.2 Measured values at
the piezoelectric buzzer
EPZ-27MS44W

Parameter	Unit	Value
Resonant frequency f_{s1}	kHz	4.457
Voltage U_A at $f = f_{s1}$	V	0.0318
Phase angle Θ at $f = f_{s1}$	Grade	25
Resonant frequency f_{p1}	kHz	4.954
Voltage U_A at $f = f_{p1}$	V	0.722
Total capacitance C at 1 kHz	nF	17.4

$$\frac{C_1}{C_0} = \left(\frac{f_{p1}}{f_{s1}}\right)^2 - 1 \tag{8.1}$$

And on the other hand it follows from Eq. (8.2) with the total capacity C the sum of these capacities.

$$C = C_1 + C_0 \tag{8.2}$$

Using the values from Table 8.2, the above equations give $C_1 = 0.23545 \cdot C_0$ and hence $C_0 = 14.08$ **nF** and $C_1 = 3.32$ **nF**. Equation (8.3) gives the series inductance L_1.

$$L_1 = \frac{1}{C_1 \cdot (\omega_{s1})^2} \tag{8.3}$$

Using the capacitance C_1 as well as $\omega_{s1} = 2 \cdot \pi \cdot f_{s1}$, calculate $L_1 = $ **384.08 mH**.

Finally, for the resonance case, the series resistance follows from Eq. (8.4) to

$$R_1 = R_V \cdot \frac{U_{R1}}{U_E - U_{R1}} \cdot \frac{1}{\cos \theta} \tag{8.4}$$

Here, U_{R1} corresponds to the voltage U_A at $f = f_{s1}$ according to Table 8.2. The calculation yields $R_1 = 362\ \Omega$. Equation (8.5) can also be used to determine the quality Q.

$$Q = \frac{1}{\omega_{s1} \cdot C_1 \cdot R_1} \tag{8.5}$$

For the buzzer EPZ-27MS44W you get $Q = 29.71$.

The series resonant frequency of the arrangement of ceramic and metal disc is determined by the material properties such as density and elasticity parameters as well as by the dimensions. From the design equation for the resonant frequency according to [2], Eq. (8.6) shows the influence of the dimensions at constant technology factor a.

$$f_{s1} = a \cdot \frac{T}{d^2} \tag{8.6}$$

With the data of the EPZ-27-MS44 buzzer according to Table 8.2, the technology factor follows to $a = 3.32$ kHz m.

Using this value, the resonant frequency for the EPZ-35 MS29 buzzer is calculated to be $f_{s1} = 3.32$ kHz m–0.56–10^{-3} m/(25–10^{-3} m)^2 = 2.975 kHz in good agreement with $f_{s1} = 2.9$ kHz according to Table 8.1.

Task: Display of the Frequency Dependence of the Output Voltage
The frequency response of the voltage U_A from Fig. 8.2 in the frequency range $\Delta f = 0.1$ to 10 kHz is to be simulated.

Analysis

- PSpice, Edit Simulation Profile
- Simulation Settings – Fig. 8.2: Analysis
- Analysis type: AC Sweep/Noise
- AC Sweep type: Logarithmic, Decade
- Start Frequency: 100 Hz
- End Frequency: 10 kHz
- Points/Decade: 1 k
- Apply: OK
- PSpice, run

The analysis result according to Fig. 8.3 largely corresponds to the measured course, in particular the values of the resonance frequencies are reproduced, see Table 8.2.

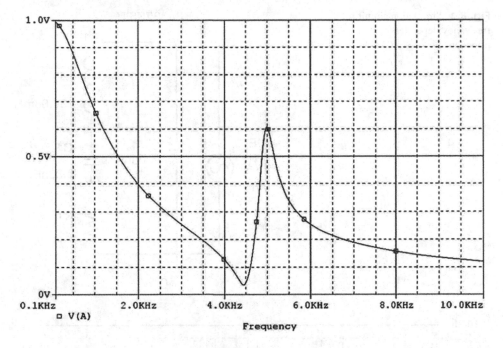

Fig. 8.3 Simulated frequency dependence of the output voltage at the buzzer

Task: Frequency Dependence of the Impedance

The circuit shown in Fig. 8.4 is to be analyzed for the frequency dependence of magnitude and phase as well as of the real and imaginary parts of the impedance of the piezoelectric buzzer EPZ-27MS44.

The frequency range shall be limited to $\Delta f = 3$ to 7 kHz.

Analysis

- PSpice, Edit Simulation Profile
- Simulation Settings – Fig. 8.4: Analysis
- Analysis type: AC Sweep/Noise
- AC Sweep type: Logarithmic/Decade
- Start Frequency: 3 kHz
- End Frequency: 7 kHz
- Points/Decade: 1 k
- Apply: OK
- PSpice, run

Figure 8.5 shows that the phase angle of the impedance Z assumes the value zero at the frequencies f_{s1} and f_{p1}. These frequencies correspond to those at which the magnitude of Z becomes minimum and maximum, respectively.

Fig. 8.4 Simulation circuit for frequency dependence of impedance

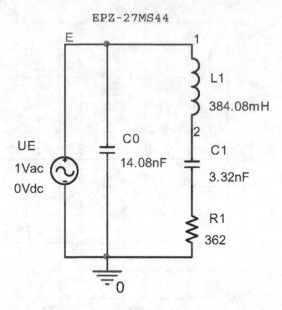

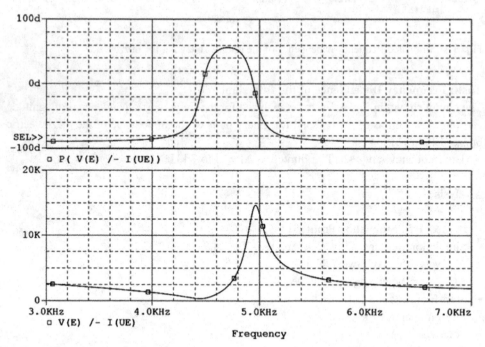

Fig. 8.5 Frequency response of magnitude and phase of the impedance of the buzzer

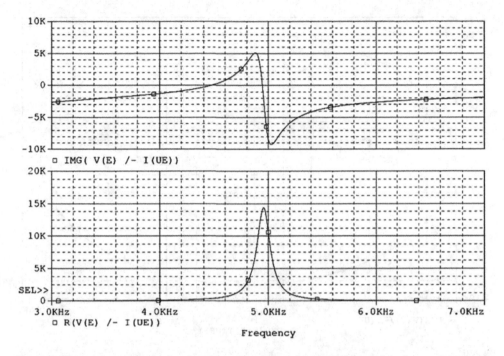

Fig. 8.6 Frequency response of real and imaginary part of the impedance of the buzzer

The frequency dependence of the real and imaginary parts is shown in Fig. 8.6. The imaginary part reaches zero at the frequency f_{s1}.

Task: Representation of the Locus Curve of the Impedance
With the circuit according to Fig. 8.4 the frequency locus of the impedance of the buzzer is to be simulated according to real and imaginary part for the range from 1 kHz to 10 kHz.

Analysis

- PSpice, Edit Simulation Profile
- Simulation Settings – Fig. 8.4
- Analysis type: AC Sweep/Noise
- AC Sweep type: Linear
- Start Frequency: 1 kHz
- End Frequency: 10 kHz
- Points/Decade: 10 k
- Apply; OK
- PSpice, run

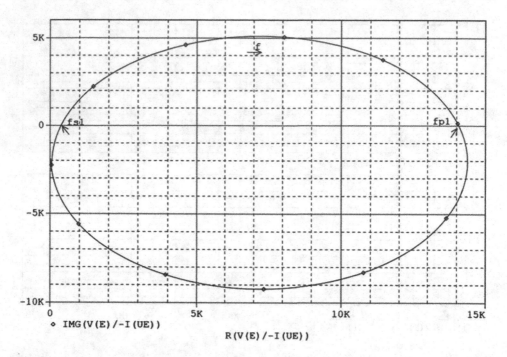

Fig. 8.7 Frequency locus of the impedance of the buzzer

Edit Diagram

- Trace, Add trace: (IMG(V(E))/-I(UE))
- Plot, Axis Settings, Axis variable
- Simulation output variable: (R(V(E)/-I(UE))
- Apply: OK
- PSpice, run

As Fig. 8.7 shows, the imaginary part is zero at frequency f_{s1} and the value of R_1 follows from $|Z|/\cos\Theta$. In the frequency range from f_{s1} up to f_{p1} the impedance of the transducer is inductive.

Buzzer EPZ-35MS29

The measurement of the output voltage across the EPZ-35MS29 buzzer results in two minima and two maxima, see Table 8.3. Thus, two C-L-R branches must be provided in the SPICE model of this converter.

The circuit for measuring the output voltage of this buzzer is shown in Fig. 8.8.

For the first C-L-R branch of this circuit, the values of the preliminary elements C_{1x}, L_{1x} as well as R_1 are determined first. After the calculation of C_2, L_2 and R_2 the provisional values are replaced by the final values for C_1 and L_1.

Table 8.3 Measured values for the piezoelectric buzzer EPZ-35MS29

Parameter	Unit	Value
Resonant frequency f_{s1}	kHz	2.981
Voltage U_A at $f = f_{s1}$	V	0.0247
Phase angle Θ at $f = f_{s1}$	Grade	0
Resonant frequency f_{p1}	kHz	3.199
Voltage U_A at $f = f_{p1}$	V	0.625
Resonant frequency f_{s2}	kHz	3.495
Voltage U_A at $f = f_{s2}$	V	0.210
Phase angle φ at $f = f_{s2}$	Grade	50
Resonant frequency f_{p2}	kHz	3.655
Voltage U_A at $f = f_{p2}$	V	0.358
Total capacity C	nF	26.1

Fig. 8.8 Circuit for measuring the AC voltage across the buzzer

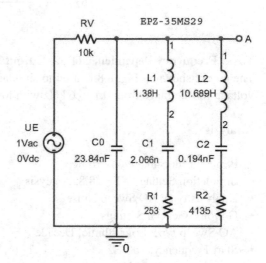

By analogy with Eq. (8.1), $C_{1x} = 0.094729 \cdot C_0$ and from Eq. (7.2) it follows $C_{1x} + C_0 = C = 26.1$ nF. From this we obtain the element values $C_0 = \mathbf{23.84\ nF}$ and $C_{1x} = 2.26$ nF.

By Eq. (8.3), $L_{1x} = 1,261$ H and Eq. (8.4) with $\Theta = 0°$ leads to $R_1 = \mathbf{253\ \Omega}$.

The capacity quotient C_2/C_{1x} is described by Eq. (8.7).

$$\frac{C_2}{C_{1x}} = \left(\frac{f_{p2}}{f_{s2}}\right)^2 - 1 \tag{8.7}$$

From this it follows that $C_2 = 0.093655 \cdot C_{1x}$.

The capacitance sum is captured by Eq. (8.8). The capacitance C_{1x} is thus divided into C_1 and C_2

$$C_{1x} = C_1 + C_2 \tag{8.8}$$

We obtain $C_1 = 2.066$ nF as well as $C_2 = 0.194$ nF and from Eq. (8.3) it follows $L_1 = 1.38$ H.

The inductance L_2 is calculated using Eq. (8.9).

$$L_2 = \frac{1}{C_2 \cdot (\omega_{s2})^2} \tag{8.9}$$

This gives $L_2 = 10.689$ H. Equation (8.10) finally gives the series resistance R_2. Here U_{R2} corresponds to the voltage U_A at $f = f_{s2}$ according to Table 8.3. $R_2 = 4135$ Ω.

$$R_2 = R_V \cdot \frac{U_{R2}}{U_E - U_{R2}} \cdot \frac{1}{cos\varphi} \tag{8.10}$$

Task: Frequency Dependence of the Output Voltage
The circuit shown in Fig. 8.8 is used to simulate the frequency dependence of the output voltage for the range from 0.1 to 6 kHz with logarithmic division of the decissa.

Analysis

- PSpice, Edit Simulation Profile
- Simulation Settings – Fig. 8.8: Analysis
- Analysis type: AC Sweep/Noise
- Options: General Settings
- AC Sweep type: Logarithmic, Decade
- Start Frequency: 100 Hz
- End Frequency: 6 kHz
- Points/Decade: 1 k
- Apply: OK
- PSpice, run

In the analysis result according to Fig. 8.9, the four resonance frequencies and the amplitudes at f_{s1} and f_{p1} of Table 8.3 are well reproduced. Deviations occur with the amplitudes to f_{s2} and f_{p2}.

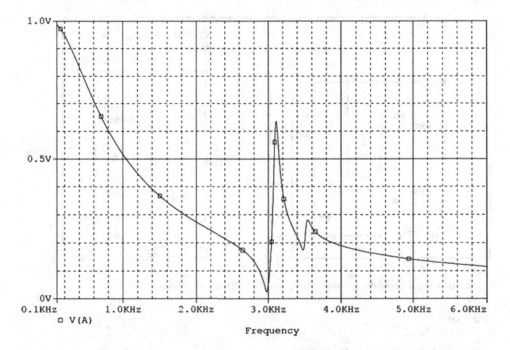

Fig. 8.9 Simulated frequency dependence of the output voltage at the buzzer

8.1.2 Circuits with Buzzers for External Control

8.1.2.1 Excitation with CMOS AMV

In the circuit shown in Fig. 8.10, a CMOS AMV drives an amplifier built with an n-channel enhancement MOSFET. The components R_S and D_1 and D_2 belong to the protection circuit. The diodes are modeled as follows:

. model DS D IS = 10f ISR = 1n CJO = 5p.

At the amplifier output, the piezoelectric buzzer is excited to sinusoidal oscillations [3, 4]. The amplitude of the sound waves is attenuated as they travel through the air to the human ear.

Task: Vibration Analysis

For the circuit shown in Fig. 8.10, the oscillations at nodes R and S in the time range $\Delta t = 0$ to 2.5 ms are to be analyzed. Furthermore, the level of the resonant frequency of the buzzer is to be determined by means of Fourier analysis.

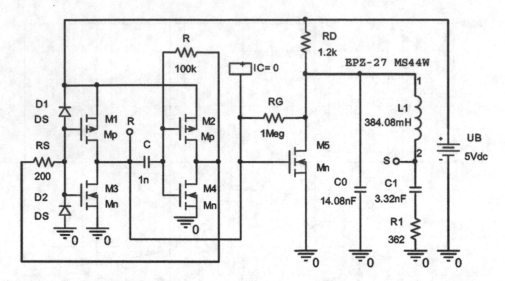

Fig. 8.10 Excitation of the buzzer by an astable CMOS multivibrator

Analysis

- PSpice, Edit Simulation Profile
- Simulation Settings – Fig. 8.10: Analysis
- Analysis type: Time Domain (Transient)
- Options: General Settings
- Run to time: 2.5 ms
- Start saving data after: 0 s
- Transient Options
- Maximum step size: 5 us
- Apply: OK
- PSpice, run

Figure 8.11 shows the output square-wave voltage of the CMOS AMV and the sinusoidal oscillations of the piezoelectric buzzer at node S.

The Fourier analysis for the frequency dependence of the voltage at node S leads to the resonant frequency at the level of 4.4 kHz, see Fig. 8.12.

8.1.2.2 RC Phase Shifter

In the circuit shown in Fig. 8.13, the buzzer is located at the output of the RC phase shifter. The necessary phase rotation of 180° is applied by the RC elements, whose voltage drop is compensated by the amplification of the inverting amplifier stage.

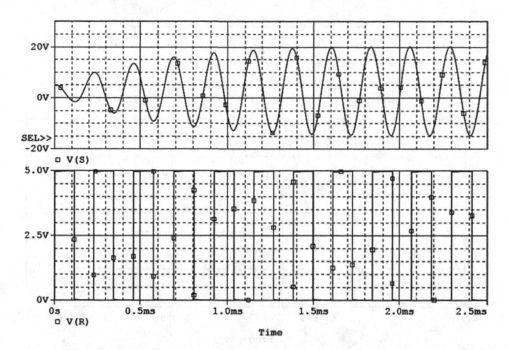

Fig. 8.11 Simulated oscillations of the multivibrator and the buzzer

Task: Analysis of Sinusoidal Oscillations

For the circuit shown in Fig. 8.13, the sinusoidal oscillations at the output of the RC phase shifter and those of the buzzer in the time range $\Delta t = 0$ to 10 ms are to be analyzed. By means of Fourier analysis, the magnitude of the oscillation frequencies is to be recorded.

Analysis

- PSpice, Edit Simulation Profile
- Simulation Settings – Fig. 8.13: Analysis
- Analysis type: Time Domain (Transient)
- Run to time: 6.0 ms
- Start saving data after: 0 s
- Maximum step size: 5 us
- Apply: OK
- PSpice, run

The buzzer receives the external control through the sinusoidal output voltage of the RC phase shifter. In Fig. 8.14 the simulated oscillations at nodes A and S are shown.

The level of the oscillation frequencies at nodes A and S is uniform at 4.53 kHz, see Fig. 8.15.

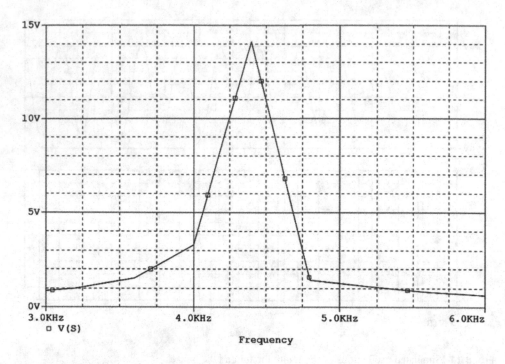

Fig. 8.12 Simulated value of the resonance frequency of the buzzer

8.1.2.3 Tapped LC Resonant Circuit

In the circuit shown in Fig. 8.16, the piezoelectric buzzer EPZ-27MS44 is excited via a resonant circuit with center tap [5]. The oscillation is made possible by setting the initial condition IC = 0.7 V. The piezoelectric buzzer is then switched on.

Task: Detection of Sinusoidal Oscillations

The oscillations of the oscillating circuit and the EPZ-27MS44 buzzer are to be represented via a transient analysis of seven to ten milliseconds. The magnitude of the oscillation frequencies shall be verified by Fourier analysis.

Analysis

- PSpice, Edit Simulation Profile
- Simulation Settings – Fig. 8.16: Analysis
- Analysis type: Time Domain (Transient)
- Run to time: 10 ms
- Start saving data after: 0 s
- Maximum step size: 5 us

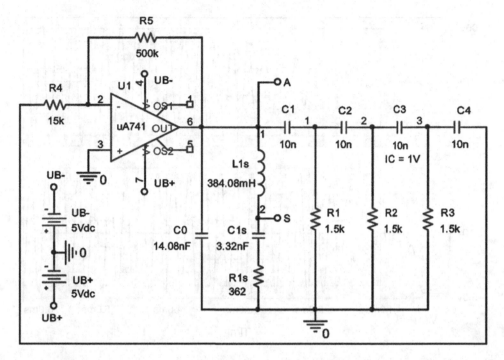

Fig. 8.13 Inclusion of the EPZ-27MS44 buzzer in an RC phase shifter

- Apply: OK
- PSpice, run

In Fig. 8.17, the oscillations at node S and between nodes L and C are shown. Figure 8.18 shows that the oscillation frequencies are at 5 kHz.

8.1.2.4 Circuit with Force Sensor

The circuit shown in Fig. 8.19 contains a multivibrator consisting of a TTL NAND gate with Schmitt trigger [6]. If a sufficiently large weight force acts on the force sensor resistor R_{FSR}, then its resistance decreases. The voltage at node E gets to H potential with values greater than 2 V and sets the multivibrator into square wave oscillations. This excites the piezoelectric buzzer EPZ-27MS44 to sinusoidal oscillations. After switching off the square-wave oscillations, the buzzer oscillations only gradually decay exponentially.

If the force sensor is not actuated, its resistance has values in the mega-ohm range and thus input E is at L potential with $V(E) \approx 0.8$ V. No oscillations occur at nodes A and S in this case.

Task: Force Sensor as On-Switch
The foil force sensor FSR 400 shall be loaded with a mass of 1 kg for a short period of time. A transient analysis over the period from 0 to 25 ms shall be used to verify the oscillations

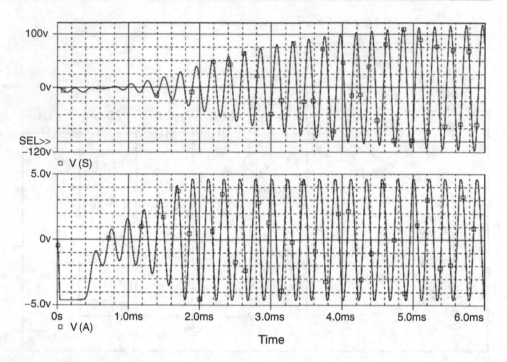

Fig. 8.14 Simulated oscillations of the RC phase shifter and the buzzer

at nodes A and S. The transient analysis shall be carried out at each node. Subsequently, the applied mass shall be reduced to fifty grams to demonstrate that no vibrations occur in this case.

Analysis

- PSpice, Edit Simulation Profile
- Simulation Settings – Fig. 8.19: Analysis
- Analysis type: Time Domain (Transient)
- Run to time: 25 ms
- Start saving data after: 0 s
- Maximum step size: 5 us
- Apply: OK
- PSpice, run

Figure 8.20 shows the square-wave oscillations of the multivibrator and the sinusoidal oscillations of the transducer when a mass of $m = 1$ kg is applied. When the mass is reduced from $m = 1$ kg to $m = 50$ g, no oscillations occur. The voltage at node A then shows the value $V(A) = 2.45$ V over the entire period and the voltage at node S is close to 0 V.

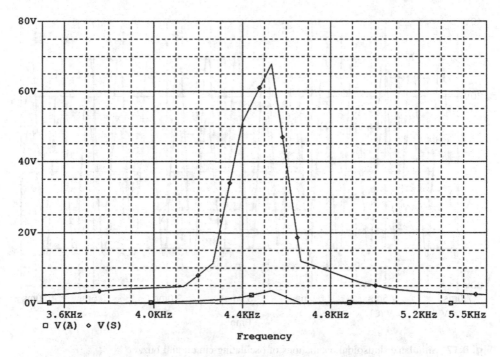

Fig. 8.15 Simulated heights of the vibration frequencies at nodes A and S

Fig. 8.16 Excitation of the buzzerr via a tapped resonant circuit

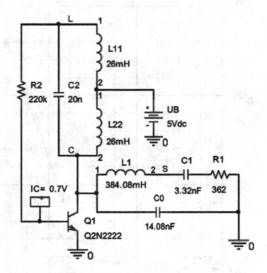

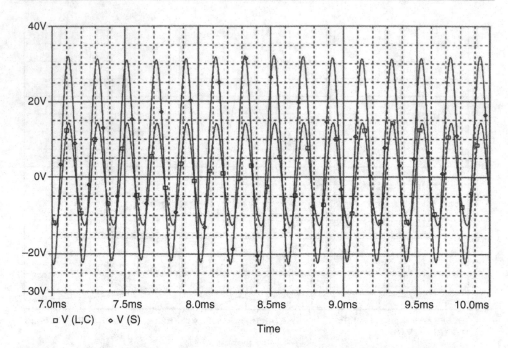

Fig. 8.17 Simulated sinusoidal oscillations of oscillating circuit and buzzer

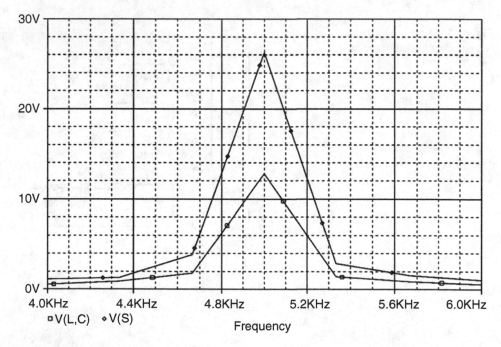

Fig. 8.18 Detection of vibration frequencies

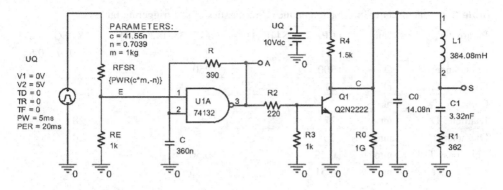

Fig. 8.19 Circuit for force sensor actuation and buzzer display

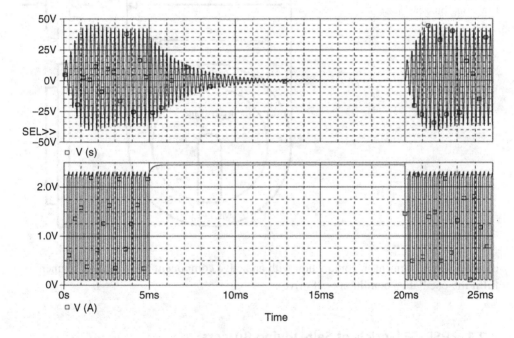

Fig. 8.20 Simulated oscillations of the multivibrator and the buzzer

8.2   Selve-Drive Buzzer

Self-drive buzzers have three electrodes: Main electrode M (MAIN), feedback electrode F
(FEEDBACK) and ground G (GROUND). In contrast to buzzers with two electrodes for
external drive, they do not require an external square-wave or sine-wave generator, but use

Table 8.4 Manufacturer information on the characteristics of self-triggering buzzers

Parameter	Unit	EPZ-27MS44F [1]	EPZ-35MS29F [1]	KPEG132 [7]
Frequency	kHz	4.4	2.9	3.0 +/− 0.5
Impedance	Ω	300	250	
Capacity C_0	nF	21	36	
Capacity C_f	nF	2.3	4.4	
Diameter D	mm	27	35	28.6
Diameter d	mm	20	25	20
Thickness T	mm	0.25	0.30	
Thickness t	mm	0.51	0.56	

Fig. 8.21 Dimensions of a self-triggering buzzer

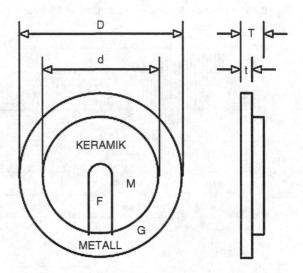

their feedback segment in conjunction with an active component and the main segment to fulfil the oscillation condition (self drive).

8.2.1   PSPICE Models of Self-Driving Buzzers

Table 8.4 shows the characteristics of self-triggering buzzers from EKULIT [1] and Kingstate Electronics Corp [7] are listed. Figure 8.21 shows a basic representation of the dimensions of the three-electrode buzzer.

Buzzer EPZ-27MS44F
The circuit shown in Fig. 8.22 is used to determine the elements of the oscillating circuit for the EPZ-27MS44F buzzer. The values entered show the final result of parameter extraction for the main and feedback segments. The values are valid for the case that the inductors L_{1M}

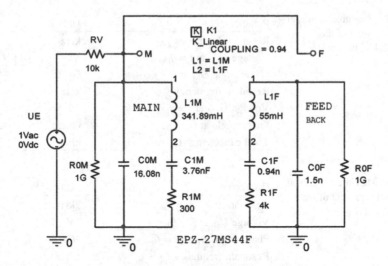

Fig. 8.22 Alternating measurement of the AC voltage at nodes M and F

and L_{1F} are coupled together. This coupling is simulated with the instruction K_Linear from the ANALOG library.

Using Eqs. (8.1), (8.2), (8.3), and (8.4), we obtain for the main segment M the capacities $C_{0M} = 16.08$ nF, $C_{1M} = 3.76$ nF, the inductance $L_{1M} = 341.89$ mH and the resistance $R_{1M} = 1039\ \Omega$. Connecting the generator to the feedback electrode F, the relation $C_F = C_{0F} + \mathbf{C}_{1F}$ holds. The two capacitances C_{0F} and C_{1F} are to be divided in conjunction with the inductance L_{1F} in such a way that the measured frequencies to the series and parallel resonance f_{sF} and f_{pF} with the associated amplitude and phase angles are approximated as well as possible. This can be achieved by reducing the resistance R_{1M} of the main segment. There is an inductive coupling. The values of the buzzer measured at the main segment M are shown in Table 8.5 and the measured values for the feedback segment F are compiled in Table 8.6.

Task: Frequency Dependence of the Alternating Voltage
For the circuit shown in Fig. 8.22, analyze the frequency dependence of the voltage by amplitude and phase in the range $\Delta f = 0$ to 10 kHz.

Analysis

- PSpice, Edit Simulation Profile
- Simulation Settings – Fig. 8.22: Analysis
- Analysis type: AC Sweep/Noise
- AC Sweep type: Linear
- Start Frequency: 1 Hz

Table 8.5 Measured values at the main segment of the EPZ-27MS44F buzzer

Parameter	Unit	Value
Resonant frequency f_{sM}	kHz	4.439
Voltage U_{sM} at $f = f_{sM}$	V	0.0731
Phase angle Θ_M at $f = f_{sM}$	Grade	−40.6
Resonant frequency f_{pM}	kHz	4.931
Voltage U_{pM} at $f = f_{pM}$	V	0.4742
Phase angle at $f = f_{pM}$	Grade	−25.7
Total capacitance C_M at 1 kHz	nF	19.84

Table 8.6 Measured values at the feedback segment of the EPZ-27MS44F buzzer

Parameter	Unit	Value
Resonant frequency f_{sF}	kHz	4.866
Voltage U_{sF} at $f = f_{sF}$	V	0.6273
Phase angle Θ_F at $f = f_{sF}$	Grade	−28
Resonant frequency f_{pF}	kHz	5.058
Voltage U_{pF} at $f = f_{pF}$	V	0.814
Phase angle at $f = f_{pF}$	Grade	−30,9
Total capacitance C_F at 1 kHz	nF	2.44

- End Frequency: 10 kHz
- Total Points: 1 k
- Apply: 1 k
- PSpice, run

The analysis results of Figs. 8.23 and 8.24 provide useful approximations to the measured values of Tables 8.5 and 8.6.

Buzzer EPZ-35MS29F
The measurement circuit for the EPZ-35MS29F buzzer is given in Fig. 8.25.
The measured values for both segments are shown in Tables 8.7 and 8.8.

Using Eqs. (8.1), (8.2), (8.3), and (8.4), obtain the element values $C_{0M} = 27.01$ nF, $C_{1M} = 3.29$ nF, $L_{1M} = 858.32$ mH, and $R_{1M} = 1148$ Ω.

Starting from $C_F = C_{0F} + C_{1F} = 3.89$ nF, a usable approximation to the measured values of Table 8.6 is obtained with the specifications $C_{0F} = 2.59$ nF and $C_{1F} = 1.30$ nF as well as $L_{1F} = 170$ mH and $R_{1F} = 4$ kΩ, if the resonant resistance of the MAIN segment is also reduced to $R_{1M} = 300$ Ω.

Task: Frequency Response of the Voltage at the Segments M and F
The circuit shown in Fig. 8.25. The frequency dependence of the amplitude and phase angle of the voltages at the main and feedback segments is to be analysed. The frequency range $\Delta f = 0$ to 6 kHz is to be recorded.

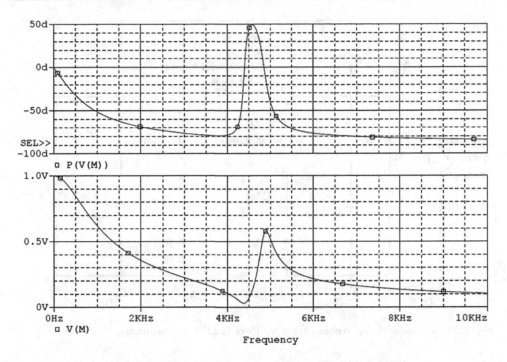

Fig. 8.23 Frequency response of amplitude and phase of the voltage at node M

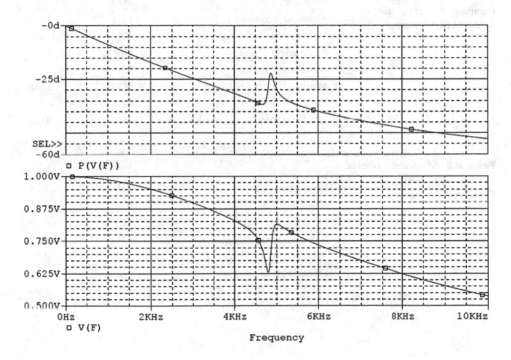

Fig. 8.24 Frequency response of amplitude and phase of the voltage at node F

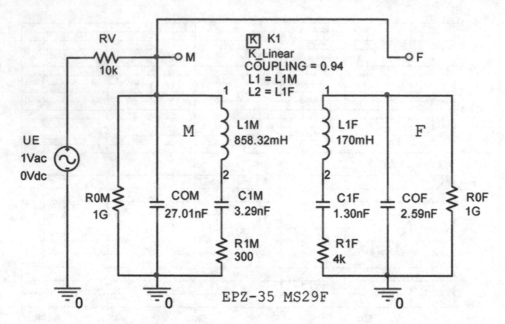

Fig. 8.25 Measuring the AC voltage at the MAIN or FEEDBACK segment

Table 8.7 Measured values at the main segment of the EPZ-35MS29F buzzer

Parameter	Unit	Value
Resonant frequency f_{sM}	kHz	2.995
Voltage U_{sM} at $f = f_{sM}$	V	0.0935
Phase angle Θ_M at $f = f_{sM}$	Grade	−26
Resonant frequency f_{pM}	kHz	3.172
Voltage U_{pM} at $f = f_{pM}$	V	0.658
Phase angle at $f = f_{pM}$	Grade	−16
Total capacitance C_M at 1 kHz	nF	30.30

Table 8.8 Measured values at the feedback segment of the EPZ-27MS44F buzzer

Parameter	Unit	Value
Resonant frequency f_{sF}	kHz	3.141
Voltage U_{sF} at $f = f_{sF}$	V	0.511
Phase angle Θ_F at $f = f_{sF}$	Grade	−19
Resonant frequency f_{pF}	kHz	3.192
Voltage U_{pF} at $f = f_{pF}$	V	0.53
Phase angle at $f = f_{pF}$	Grade	−22
Total capacitance C_F at 1 kHz	nF	3.89

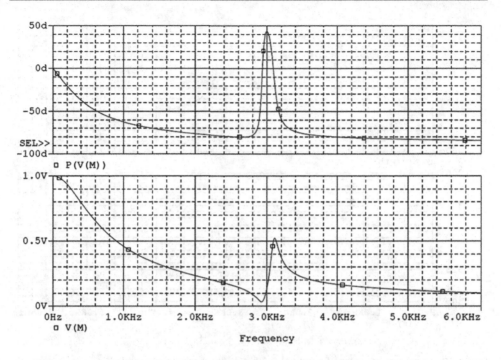

Fig. 8.26 Frequency response of amplitude and phase of the voltage at node M

Analysis

- PSpice, Edit Simulation Profile
- Simulation Settings – Fig. 8.25: Analysis
- Analysis type: AC Sweep/Noise
- AC Sweep type: Linear
- Start Frequency: 1 Hz
- End Frequency: 6 kHz
- Total Points: 1 k
- Apply: 1 k
- PSpice, run

Figure 8.26 provides an approximation to the measured resonance frequencies, voltage amplitudes and phase angles of the main segment M according to Table 8.7. For the feedback segment F, a usable agreement of the simulation values according to Fig. 8.27 with the measured values according to Table 8.8 is obtained.

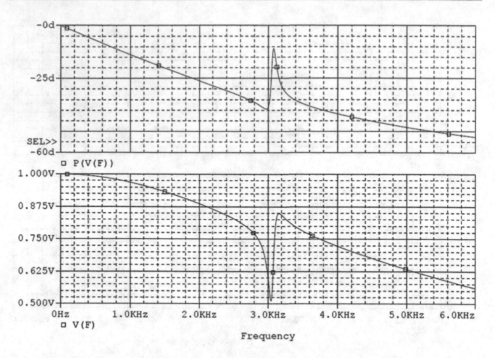

Fig. 8.27 Frequency response of amplitude and phase of the voltage at node F

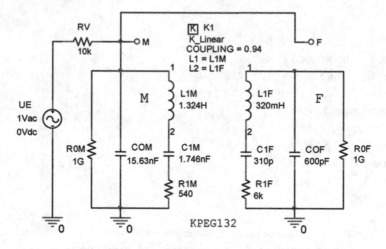

Fig. 8.28 Measuring circuit for recording the voltages at nodes M and F

Buzzer KPEG132

The measuring circuit for this self-triggering buzzer in a plastic housing with a sound aperture is shown in Fig. 8.28. Note the inductive coupling between the elements L_{1M} and L_{1F}, which is implemented with the K_Linear module from the analog library.

Table 8.9 Measured values at the main segment of the EPG132 buzzer

Parameter	Unit	Value
Resonant frequency f_{sM}	kHz	3.31
Voltage U_{sM} at $f = f_{sM}$	V	0.0543
Phase angle Θ_M at $f = f_{sM}$	Grade	−20
Resonant frequency f_{pM}	kHz	3.49
Voltage U_{pM} at $f = f_{pM}$	V	0.673
Phase angle at $f = f_{pM}$	Grade	−15.9
Total capacitance C_M at 1 kHz	nF	17.38

Table 8.10 Measured values at the feedback segment of buzzer EPG132

Parameter	Unit	Value
Resonant frequency f_{sF}	kHz	3.61
Voltage U_{sF} at $f = f_{sF}$	V	0.8616
Phase angle Θ_F at $f = f_{sF}$	Grade	−16.8
Resonant frequency f_{pF}	kHz	3.723
Voltage U_{pF} at $f = f_{pF}$	V	0.953
Phase angle at $f = f_{pF}$	Grade	−16.7
Total capacitance C_F at 1 kHz	nF	0.91

The measured values at the main and feedback segments are summarized in Tables 8.9 and 8.10.

Using Eqs. (8.1), (8.2), (8.3), and (8.4), calculate $C_{0M} = 15.63$ nF, $C_{1M} = 1.746$ nF, $L_{1M} = 1.324$ H, and $R_{1M} = 611\ \Omega$. It holds that $C_F = C_{0F} + C_{1F} = 0.91$ nF.

From the division with $C_{0F} = 600$ pF and $C_{1F} = 310$ pF as well as with $L_{1F} = 320$ mH and $R_{1F} = 6$ kΩ, good approximations to the measured values of Table 8.10 are ultimately obtained if, furthermore, the resonant resistance of the main segment is reduced to $R_{1M} = 540\ \Omega$.

Task: Frequency Dependence of the Voltage at the Nodes M and F

For the circuit shown in Fig. 8.28, analyze the frequency dependence of the voltage on the main and feedback segments in the range $\Delta f = 0$ to 6 kHz.

Analysis

- PSpice, Edit Simulation Profile
- Simulation Settings – Fig. 8.28: Analysis
- Analysis type: AC Sweep/Noise
- AC Sweep type: Linear
- Start Frequency: 1 Hz
- End Frequency: 6 kHz

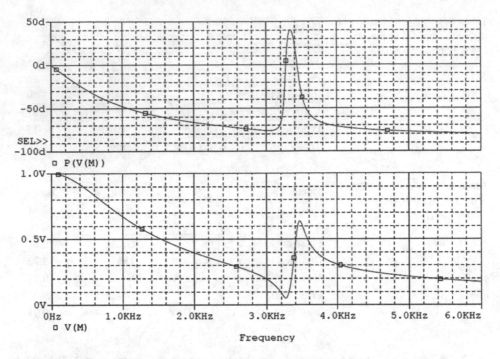

Fig. 8.29 Frequency dependence of the voltage at the main segment

- Total Points: 1 k
- Apply: 1 k
- PSpice, run

The analysis results for both buzzer segments according to Figs. 8.29 and 8.30 agree well with the corresponding measured values of Tables 8.9 and 8.10.

8.2.2 Circuits with Self-Triggering Buzzers

In the following, the usefulness of the buzzer models is demonstrated for the three types studied, EPZ-27MS44F, EPZ-35MS29F and EPG132.

8.2.2.1 Transistor Oscillator

Figure 8.31 presents a self-triggering circuit [8] with a bipolar transistor as the active component. The electrode M is at the output and the electrode F at the input of the circuit. This achieves the required phase position.

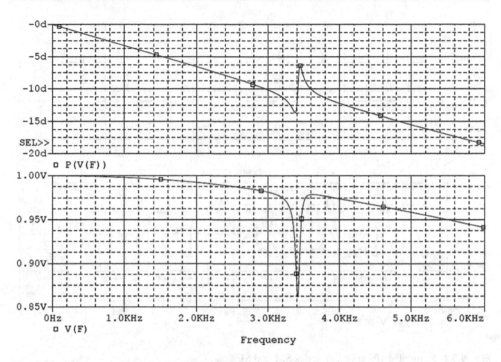

Fig. 8.30 Frequency dependence of the voltage at the feedback segment

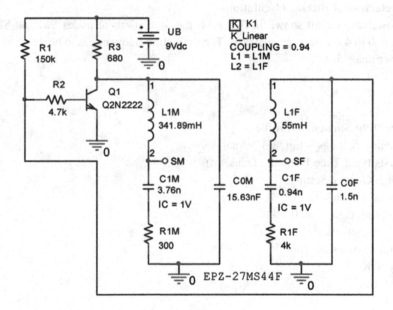

Fig. 8.31 Transistor oscillator with buzzer EPZ-27MS44F

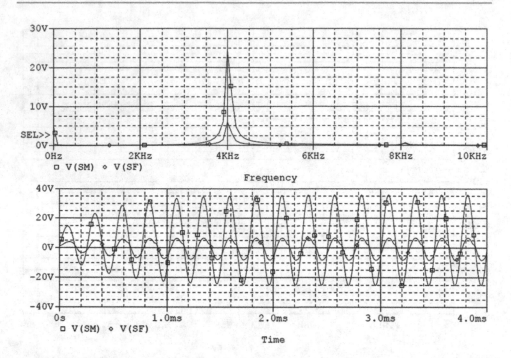

Fig. 8.32 Simulated vibrations at nodes SM and SF

Task: Detection of Buzzer Oscillations

For the oscillator circuit shown in Fig. 8.31, the oscillations at nodes SM and SF in the range $\Delta t = 0$ to 4 ms are to be analyzed. The oscillation frequency is to be determined with the Fourier analysis.

Analysis

- PSpice, Edit Simulation Profile
- Simulation Settings – Fig. 8.31: Analysis
- Analysis type: Time Domain (Transient)
- Options: General Settings
- Run to time: 4 ms
- Start saving data after: 0 s
- Transient Options
- Maximum step size: 1 us
- Apply, OK
- PSpice, run

The analysis first provides the time dependence of the voltages with Fig. 8.32. Then the time axis is converted into the frequency axis with Plot, Add Plot to Window, Plot,

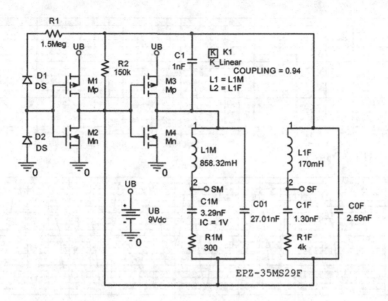

Fig. 8.33 CMOS inverter oscillator

Unsynchronous Axis, Plot, Axis Settings, Fourier, Add Trace, Trace Expression: V(SM), V (SF). The frequency range can be limited to 10 kHz via User defined. The Fourier analysis yields the oscillation frequency with $f_0 = 4$ kHz.

8.2.2.2 CMOS Inverter Oscillator

Figure 8.33 shows a circuit according to [8, 9], which is constructed with two CMOS inverters. For the circuit CD4007 the following modeling is applied:

. model Mn NMOS L = 5u W = 100u KP = 30u LAMBDA = 10 m
. model Mp PMOS L = 5u W = 200u KP = 15u LAMBDA = 10 m
. model DS D IS = 10f ISR = 1n CJO = 10p.

Task: Detection of Vibrations at Nodes SM and SF

For the circuit shown in Fig. 8.33, the time domain analysis for $\Delta t = 0$ to 4 ms is to be carried out. The oscillation curve at nodes SM and SF is to be displayed. The magnitude of the oscillation frequency is to be determined with the Fourier analysis.

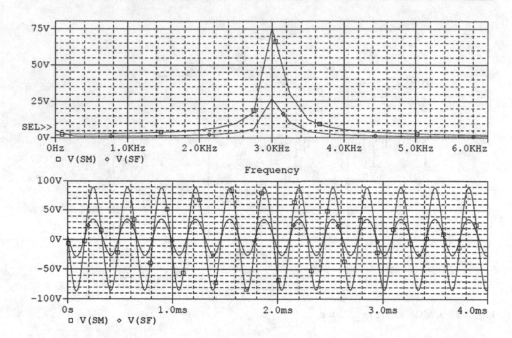

Fig. 8.34 Vibration curve at buzzer EPZ-35MS2

Analysis

- PSpice, Edit Simulation Profile
- Simulation Settings – Fig. 8.33: Analysis
- Analysis type: Time Domain (Transient)
- Options: General Settings
- Run to time: 4 ms
- Start saving data after: 0 s
- Transient Options
- Maximum step size: 1 us
- Apply, OK
- PSpice, run

The analysis result according to Fig. 8.34 shows the vibration frequency with $f_0 = 3$ kHz in good agreement with the data sheet specification.

The oscillation amplitudes at the main segment MAIN are significantly higher than those of the feedback segment FEEDBACK.

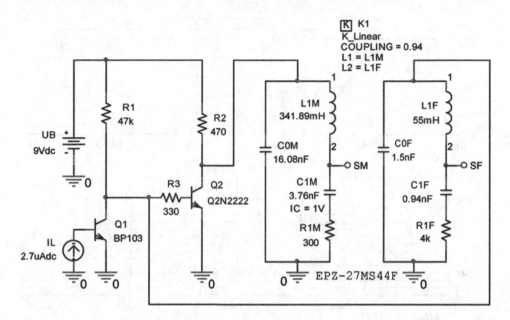

Fig. 8.35 Circuit with phototransistor

8.2.2.3 Circuit with Phototransistor

In the circuit shown in Fig. 8.35, the buzzer EPZ-27MS44F sounds an alarm when the phototransistor does not receive light and thus transistor Q_2 is activated.

Task: Effect of a Lighting

For the circuit shown in Fig. 8.35, the effects are to be investigated for the cases where the phototransistor is darkened with $I_L = 0$ μA on the one hand and illuminated with illuminance $E_v = 1$ klx corresponding to $I_L = 2.7$ μA on the other hand. The phototransistor is modeled via Edit, Pspice Model as follows:

.model BP103 NPN IS = 10f BF = 352 VAF = 360

Analysis

- PSpice, Edit Simulation Profile
- Simulation Settings – Fig. 8.35: Analysis
- Analysis type: Time Domain (Transient)
- Options: General Settings
- Run to time: 4 ms, 6 ms
- Start saving data after: 0 s
- Transient Options

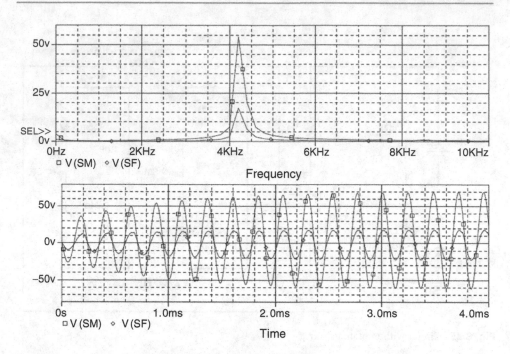

Fig. 8.36 Oscillations at the buzzer with unlit phototransistor

- Maximum step size: 1 us
- Options: Parametric Sweep
- Sweep variable: Global Parameter
- Parameter Name: IL
- Sweep type: Value list: 0 2.7 uA
- Apply, OK
- PSpice, run

Figure 8.36 shows the oscillations of the self-drive buzzer EPZ-27MS44F in the case that the phototransistor does not receive any light. The oscillation frequency $f_0 = 4.4$ kHz is identical to the data sheet specification.

When the phototransistor is illuminated with $E_v = 1$ klx, it switches on. This means that only a small saturation voltage is present across its collector, which means that transistor Q_2 is switched off and the buzzer remains silent after a short time, see Fig. 8.37.

8.2.2.4 Circuit with NTC Sensor

In the circuit according to Fig. 8.38, the NTC sensor M87–10 reaches the value $R = 10$ kΩ at $T = 298$ K corresponding to 25 °C. As the temperature decreases, the sensor resistance increases. Consequently, for $T < 298$ K, $U_P > U_N$. Thus, the output voltage of the

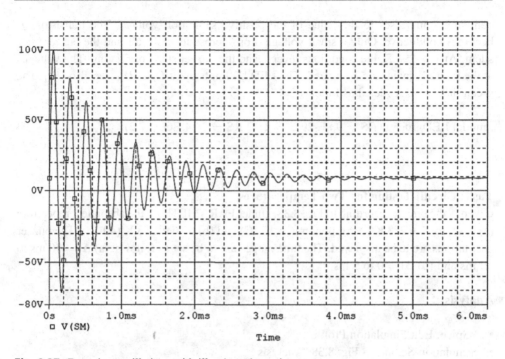

Fig. 8.37 Decaying oscillations with illuminated transistor

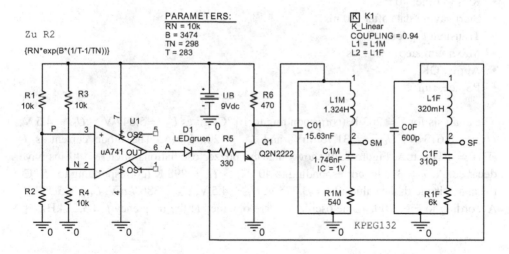

Fig. 8.38 Circuit with NTC sensor

comparator increases to the value of the positive saturation voltage. The transistor Q_1 becomes conductive. Only a small voltage drops across its collector and the buzzer goes silent. At $T > 298$ K, the conditions reverse and the buzzer starts oscillating. The voltages at the P and N inputs are compared. The LED D_1 lights up at $U_P > U_N$, see also [10]. The green lit LED is modeled as follows:

. model LEDgruen D (IS = 9.8E-29 N = 1.12 RS = 24.4 EG = 2.2).

Task: Circuit Testing for Two Temperatures
Given is the circuit according to Fig. 8.38. Under PARAMETERS the absolute temperature (a) is set to $T = 283$ K (corresponding to 10 °C) and (b) to $T = 313$ K (40 °C). The buzzer oscillations are to be displayed. The time ranges are to be set as follows: (a) $\Delta t = 10$ ms to 40 ms and (b) $\Delta t = 0$ to 5 ms.

Analysis

- PSpice, Edit Simulation Profile
- Simulation Settings – Fig. 8.38: Analysis
- Analysis type: Time Domain (Transient)
- Options: General Settings
- Run to time: 40 ms
- Start saving data after: 10 ms
- Transient Options
- Maximum step size: 1 us
- Apply, OK
- PSpice, run

The analysis for $T = 283$ K (corresponding to 10 °C) gives $U_P = 5.847$ V $> U_N = 4.5$ V, $U_A = 8.613$ V, $U_{CE} = 31.75$ mV, and $I_C = 20.87$ mA. The diode current is I-$(D_1) = 17.08$ mA. Figure 8.39 shows that the buzzer oscillations after $t = 40$ ms have decayed. At $T = 313$ K (corresponding to 40 °C) $> T_N = 298$ K (corresponding to 25 °C) the analysis yields the values: $U_P = 3.275$ V, $U_N = 4.5$ V, $U_A = 386.7$ mV, $U_{CE} = 5.277$ V. According to Fig. 8.40, stable oscillations are obtained at the frequency $f = 3.22$ kHz.

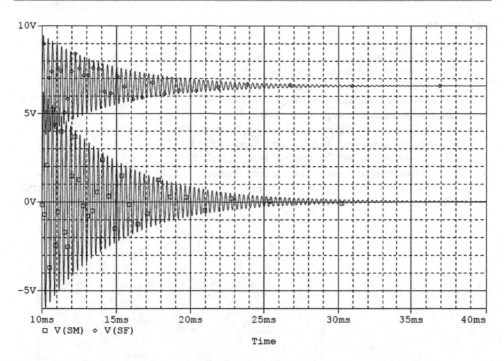

Fig. 8.39 Decaying oscillations at the sensor temperature of 10 °C

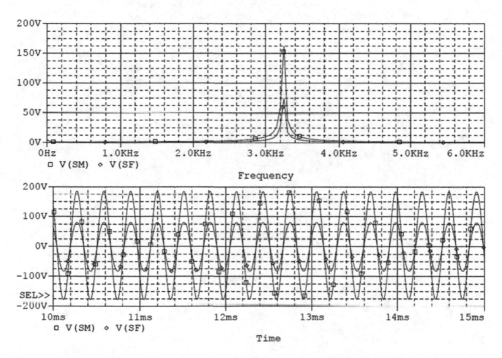

Fig. 8.40 Oscillations of the buzzer at the sensor temperature of 40 °C

References

1. EKULIT: Datenblätter zu piezoelektrischen Schallwandlern, Ostfildern/Nellingen (2014)
2. APC International: Piezo Buzzers, Firmenschrift (2013)
3. Kyocera: Piezoelectric Acoustic Generators, Application Report (2014)
4. http://www.microbuzzer.com/magnetic-piezo-buzzer-circuits
5. http://www.b.kainka.de/kosmos/summer.htm
6. Weddigen, C., Jüngst, W.: Elektronik. Springer, Berlin (1993)
7. Kingstate Electronics Corp.: Datenblatt zum Summer KPEG132 (2017)
8. Murata: Piezoelectric Sound Components, Firmenschrift (2014)
9. EKULIT: Introduction of Piezoelectric Ceramic Element and Buzzer, Firmenschrift (2015)
10. Hareendran, T.K.:Shadow Sensor Alarm. Electro Schematics. http://www.electroschematics.com/8386/shadow-sensor-alarm/ (2017)

Ultrasonic Transducer

9

Ultrasonic transducers operate at frequencies from 20 kHz to several gigahertz. They are used both to emit sound waves and to receive them. The following embodiments are limited to ultrasonic transducers with an operating frequency of 40 kHz.

9.1   PSPICE Models

From Eqs. (8.1), (8.2), (8.3) and (8.4) of the previous chapter and the transducer characteristics according to Table 9.1, the L-C-R parameters of the transmitter and receiver follow according to Table 9.2.

Data Sheet
Table 9.1 shows the characteristics of a US transmitter and a US receiver from Pro-Wave Electronics Corp. The series and parallel resonance frequencies as well as the phase angle of the magnitude of the impedance at series resonance were taken from the diagrams provided by the manufacturer.

Task: Frequency Dependence of the Impedance
For the US transmitter and receiver shown in Fig. 9.1, the frequency dependence of the impedances Z_T and Z_R and the phase angles P_T and P_R in the frequency range is as follows $\Delta f = 35$ to 45 kHz.

© The Author(s), under exclusive license to Springer Fachmedien Wiesbaden GmbH, part of Springer Nature 2023
P. Baumann, *Selected Sensor Circuits*,
https://doi.org/10.1007/978-3-658-38212-4_9

Table 9.1 Characteristics of ultrasonic transducers according to [1]

US converter	Transmitter 400ST120	Receiver 400SR120
Frequency	40 +/− 1 kHz	40 +/− 1 kHz
Transmitter sound pressure at 40 kHz or 0.0002 μbar for 10 V_{eff} at 30 cm	115 dB min.	–
Receiver sensitivity at 40 kHz, 0 dB or 1 V/μbar	–	−67 dB min.
Capacity at 1 kHz	2400 pF	2400 pF
Series resonant frequency f_s	40.24 kHz	38.90 kHz
Parallel resonant frequency f_p	41.50 kHz	40.25 kHz
Amount of impedance $|Z|$ at $f = f_s$	530 Ω	630 Ω
Phase angle Θ of $|Z|$ at $f = f_s$	−30°	−40°

Table 9.2 Model parameters of 400ST/R120 series ultrasonic transducers

US converter	Transmitter 400ST120	Receiver 400SR120
Series inductance L_1	108.63 mH	105.95 mH
Series capacity C_1	144 pF	158 pF
Series resistance R_1	612 Ω	822 Ω
Parallel capacitance C_0	2256 pF	2242 pF

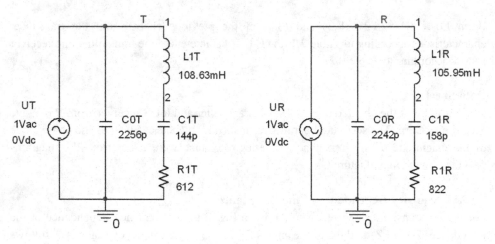

Fig. 9.1 Circuits for simulating US transmitter and receiver

Analysis

- PSpice, Edit Simulation Profile
- Simulation Settings – Fig. 9.1: Analysis
- Analysis type: AC Sweep/Noise

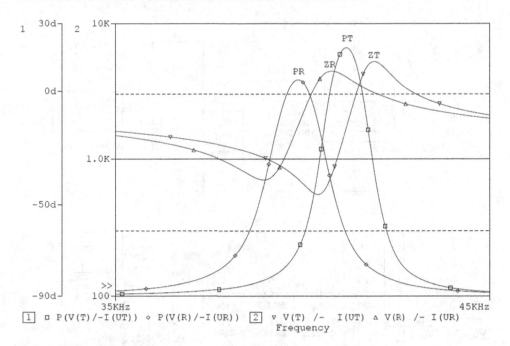

Fig. 9.2 Frequency dependence of impedance and phase angle of transmitter and receiver

- Options: General Settings
- AC Sweep type: Linear
- Start Frequency: 35 kHz
- End Frequency: 45 kHz
- Total Points: 1 k
- Apply: OK
- PSpice, run

The analysis result according to Fig. 9.2 largely corresponds to the manufacturer's diagram of the ultrasonic transducers under consideration with regard to the frequency dependence of impedance and phase. Thus the series and parallel resonance frequencies are also achieved.

Table 9.2 shows the extracted model parameters of the US transmitter and US receiver. There are only minor differences in the values.

9.2   Ultrasonic Transmitter and Receiver

Based on the determined models of the US transmitter and receiver, transmitter and receiver circuits for a transmission of sound waves through the medium air can be analyzed with the program PSPICE.

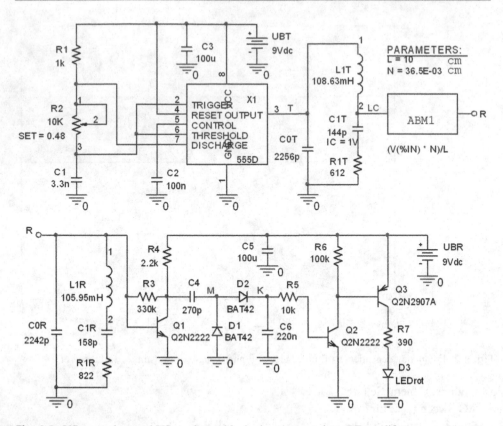

Fig. 9.3 US transmitter and US receiver with single-stage transistor RF amplifier

9.2.1   US Transmitter and US Receiver as Single Stage Transistor Amplifier

The circuit shown in Fig. 9.3 is designed so that the LED just lights up at the length $L = 10$ cm between the transmitter and receiver. The transmitter is implemented with the 555D timer circuit and the ABM1 component from the ABM library causes the sine signal U_{LC} to attenuate as the length L increases.

Task: Signal Transmission
For the time interval $\Delta t = 1.9$ to 2 ms, the stresses at nodes T, LC, and R are to be analyzed.

Analysis

- PSpice, Edit Simulation Profile
- Simulation Settings – Fig. 9.3: Analysis
- Analysis type: Time Domain (Transient)
- Run to time: 2 ms

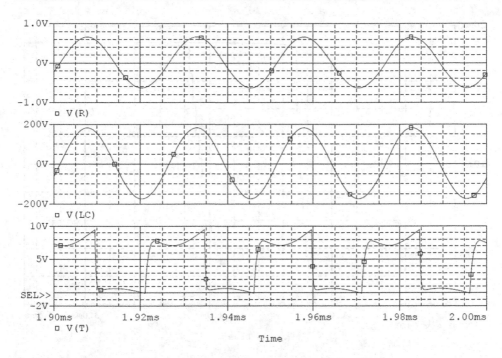

Fig. 9.4 Transmitter pulses and transmitted and received sinusoidal oscillations

- Start saving data after: 1.9 ms
- Maximum step size: 0.5 us
- Apply: OK
- PSpice, run

The analysis result according to Fig. 9.4 shows the positive square-wave pulses with the pulse frequency $f = 40$ kHz applied to the transmitter T, furthermore the sinusoidal sound waves radiated with relatively high amplitude at the node LC as well as the sinusoidal waves of smaller amplitude appearing at the receiver R. The analysis result according to Fig. 9.4 shows the positive square-wave pulses with the pulse frequency $f = 40$ kHz radiated at the node LC.

Task: LED Current with Variation of the Length Between Transmitter and Receiver
Vary the length L in the circuit shown in Fig. 9.3 as 8, 10 and 12 cm and analyze the effect on the LED current in the time range $\Delta t = 0$ to 3.5 ms.

Analysis

- PSpice, Edit Simulation Profile
- Simulation Settings – Fig. 9.3: Analysis

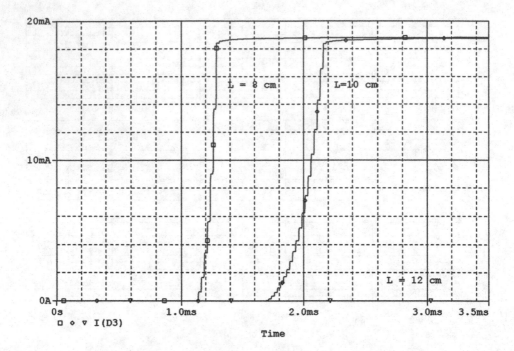

Fig. 9.5 LED currents at different distances of the receiver from the transmitter

- Analysis type: Time Domain (Transient)
- Options: General Settings
- Run to time: 3.5 ms
- Start saving data after: 0 s
- Maximum step size: 0.5 us
- Sweep variable: Global Parameter
- Parameter Name: L
- Sweep type: value list: 8 10 12
- Apply: OK
- PSpice, run

From Fig. 9.5 it can be seen that the LED lights up at $L \leqq 10$ cm, but is already inactive at $L = 12$ cm.

9.2.2 US Transmitter with US Receiver as OP- NF Amplifier

The circuit shown in Fig. 9.6 again contains the US transmitter with the 555D circuit as well as an LF amplifier for which an operational amplifier with a high-impedance input is to be used. This OP of type LF411 is only supplied with the positive operating voltage [2, 3].

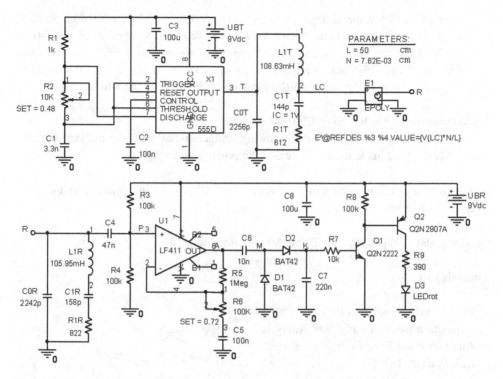

Fig. 9.6 US transmitter and OP-NF amplifier as receiver

For a length L lying between the US transmitter and the US receiver, the voltage received at node R can be estimated using the steps given by Pro-Wave Electronics Corp. in [4].

With the default $L = 50$ cm and the standard length $L_S = 30$ cm, the transmitter voltage $V(T) = 6.59$ V_{eff}, the standard transmitter voltage $V(T_S) = 10$ V_{eff} and the data of the US transducers used according to Table 9.1, the following values are obtained for the transmitter sound pressure level (*SPL*) and the receiver sensitivity (*S*):

1. Reduction of *SPL* = 115 dB with respect to V(T): $20\text{-}\log(V(T)/V(Ts)) = -3.62$ dB
2. Reduction of *SPL* with respect to L: $20\text{-}\log(L_S/L) = -4.44$ dB
3. Reduction of *SPL* with respect to absorption in air: 0.1886 dB/m-L = -0.09 dB
4. Result: $SPL = 115$ dB $- 8.15$ dB $= 106.85$ dB, $10SPL/20 = 10^{5.3425} = 220039$
5. Conversion of *SPL* to µbar: $X = 10SPL/20 - 0.0002$ µbar = 44.01 µbar resp.
6. 106.85 dB = $20\text{-}\log(X/0.0002$ µbar), X = 44.01 µbar
7. Receiver $S = -67$ dB = $20\text{-}\log(Y/1$ V/µbar), $Y = 10^{S/20} = 10^{-3.35} = 0.44668$ mV/µbar
8. Receiver: V(R) = X·Y = 19.66 mV

For the source EPOLY the voltage quotient is $V(R)/V(LC) = GAIN = 19.66 \text{ mV}_{\text{eff}}/129$ $V_{\text{eff}} = 1.524–10^{-4}$. The value of voltage V(LC) was recorded as the result of simulation.

From $VALUE = V(R) = V(LC)\text{-}N/L = 19.66$ mV follows for $L = 50$ cm the value $N = 7.62–10^{-3}$ cm. This value is approximately constant in the range $L = 30$ to 100 cm. With the doubling of L, the voltage V(R) received at the receiver then drops to half.

Task: Analysis of US Receiver Characteristics

For the circuit shown in Fig. 9.6, the following characteristics are to be analysed with the setting SET $= 0.72$ made for $L = 50$ cm at the potentiometer R_6:

1. The rms value of the voltage at node R and the LED current in the time domain
2. $\Delta t = 0$ to 15 ms.

the sinusoidal voltages at nodes P and A in the time range $\Delta t = 14.9$ to 15 ms.

Analysis to 1

- PSpice, Edit Simulation Profile
- Simulation Settings – Fig. 9.6: Analysis
- Analysis type: Time Domain (Transient)
- Run to time: 15 ms
- Start saving data after: 0
- Maximum step size: 0.5 us
- Apply: OK
- PSpice, run

The analysis yields with Fig. 9.7 the previously calculated receiver voltage V (R) $= 19.66$ mV. With the setting SET $= 0.72$ at the potentiometer R_6 the LED just becomes active.

Analysis to 2

- PSpice, Edit Simulation Profile
- Simulation Settings – Fig. 9.6: Analysis
- Analysis type: Time Domain (Transient)
- Run to time: 15 ms
- Start saving data after: 14.9 ms
- Maximum step size: 0.5 us
- Apply: OK
- PSpice, run

The analysis result according to Fig. 9.8 provides the OP input voltage with $U_{\text{Pss}} = 0.0552$ V and the OP output voltage $U_{\text{Ass}} = 0.8212$ V. This gives an ac gain $v_{\text{u}} = 14.88$. Given the

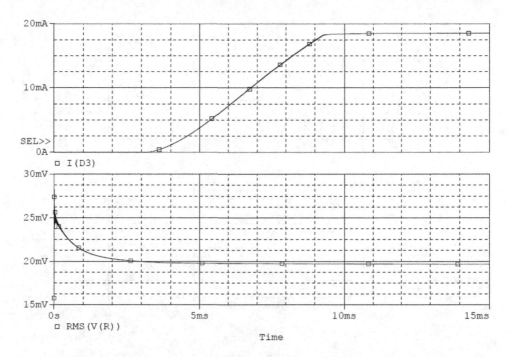

Fig. 9.7 RMS value of the voltage at the receiver and onset of LED current

negligible value of the reactance of capacitor C_5 of about 40 ohms, Eq. (9.1) holds. We obtain $v_u = 1 + 1$ MΩ/72 kΩ = 14.89.

$$v_u = 1 + \frac{R_5}{R_6} \tag{9.1}$$

Task: Variation of the Length Distance Between Transmitter and Receiver

In the circuit shown in Fig. 9.6, make the following changes: $L = 100$ cm and $SET = 0.30$. With the variation of the length distance $L = 50$, 70 and 100 cm, the RMS value of the receiver voltage and the LED current in the time range $\Delta t = 0$ to 5 ms are to be analyzed.

Analysis

- PSpice, Edit Simulation Profile
- Simulation Settings – Fig. 9.6: Analysis,
- $L = 100$, SET = 0.30
- Analysis type: Time Domain (Transient)
- Run to time: 5 ms
- Start saving data after: 0
- Transient Options

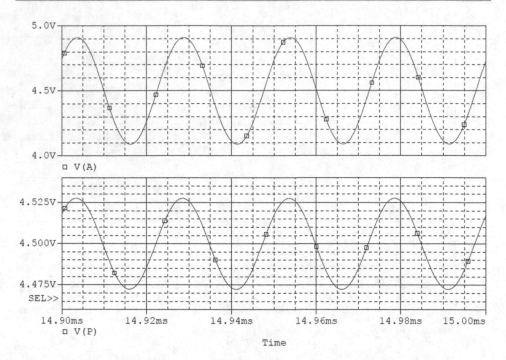

Fig. 9.8 Input and output voltage at the non-inverting operational amplifier

- Maximum step size: 0.5 us
- Options: Parametric Sweep
- Sweep variable: Global Parameter
- Parameter Name: L
- Sweep type: Value list: 50, 70, 100
- Apply: OK
- PSpice, run

The analysis shows that the rms value of the voltage at the receiver drops to half when the length L is doubled. The onset of the LED current is delayed with increasing length, see (Fig. 9.9).

9.3   Ultrasonic Distance Warning

If the transmitter component is included in the original receiver circuit and excited to oscillate, then acoustic feedback can occur when the transmitter-receiver arrangement approaches a reflector wall. Such a circuit can be used as a distance warning device in the sense of an ultrasonic barrier [5, 6]. The required excitation, which in practice can be generated via a noise, is realized in the simulation via an initial condition IC (Initial

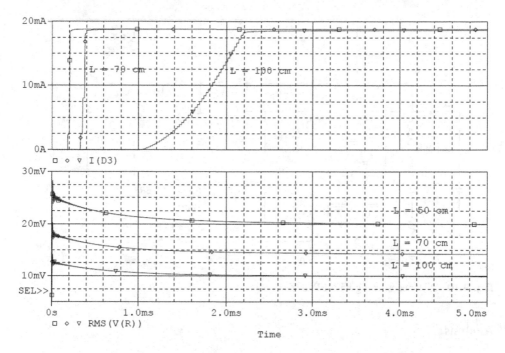

Fig. 9.9 Receiver voltage and LED current when the transmitter approaches

Condition) at the capacitor C_{1T}. The attenuation of the emitted ultrasonic sine waves is done by an E-source as a coupling element between transmitter and receiver. Another coupling between transmitter and receiver can be implemented in PSPICE with the K_linear module from the ANALOG library with regard to the transducer inductances.

9.3.1 Distance Warning with E-POLY Source as Coupling Element

The circuit of the ultrasonic distance warning device with the voltage-controlled voltage source E-POLY between the transmitter tap LC and the receiver R is shown in Fig. 9.10.

Task: Dependence of the LED Current on the GAIN Value
With an analysis in the time interval $\Delta t = 0$ to 2.5 ms, the minimum value of GAIN at which the LED just lights up with increasing distance between both US baffles and the reflector wall shall be determined. Furthermore, it has to be proven with which GAIN values the LED is active. Because with the simple E-source the parameter GAIN cannot be varied, the E-POLY source is equipped with a parameter instruction for GAIN. The following applies: GAIN = V(R)/V(LC) and VALUE = V(LC)*GAIN = V(R).

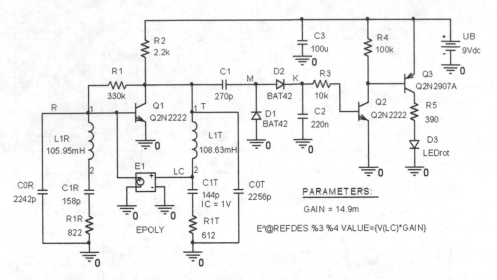

Fig. 9.10 Distance warning with EPOLY source

Analysis

- PSpice, Edit Simulation Profile
- Simulation Settings – Fig. 9.10: Analysis,
- Analysis type: Time Domain (Transient)
- Run to time: 2.5 ms
- Start saving data after: 0
- Transient Options
- Maximum step size: 0.2 us
- Options: Parametric Sweep
- Sweep variable: Global Parameter
- Parameter Name: GAIN
- Sweep type: Value list: 14 m 14.9 m 25 m 50 m
- Apply: OK
- PSpice, run

The analysis provides the minimum value at which the LED is still lit, with GAIN $= 14.9$–10^{-3}. This value applies to the measured maximum distance $A = 22$ cm between the two US transducers located at the same height and the reflector wall. As Fig. 9.11 shows, the LED is no longer active at GAIN $= 14$ m, but it is consistently illuminated at GAIN ≥ 14.9 m. The coupling in the circuit of Fig. 9.10 via the GAIN parameter of the voltage-coupled voltage source EPOLY is subsequently replaced by inductive coupling via the K_linear parameter in the circuit of Fig. 9.14.

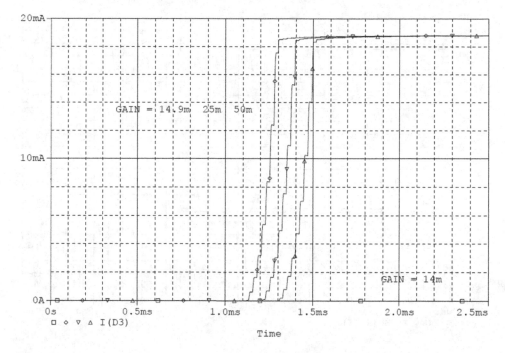

Fig. 9.11 Dependence of the LED current on the GAIN parameter

Task: AC Voltages at Transmitter and Receiver

With the parameter GAIN = $14.9–10^{-3}$ of the circuit shown in Fig. 9.10, the AC voltages at the transmitter T and receiver R in the time range Δt = 2.2 to 2.5 ms to analyze.

Analysis

- PSpice, Edit Simulation Profile
- Simulation Settings – Fig. 9.10: Analysis,
- Analysis type: Time Domain (Transient)
- Run to time: 2.5 ms
- Start saving data after: 2.2 ms
- Transient Options
- Maximum step size: 0.2 us
- Apply: OK
- PSpice, run

The result of the analysis is a superimposed alternating voltage of a few volts at the transmitter. From the tap at node LC a sinusoidal voltage reaches the input of the E-source. The sinusoidal voltage reduced with GAIN reaches the receiver. The rms value

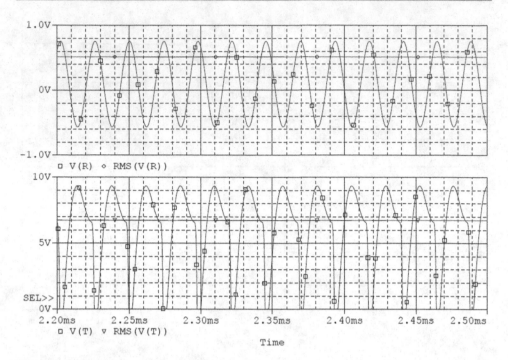

Fig. 9.12 AC voltages at transmitter and receiver

of the sinusoidal voltage at the receiver is about 0.5 V at the base-emitter junction of transistor Q_1, see Fig. 9.12.

Task: Stresses at Nodes M and K
For the circuit shown in Fig. 9.10 with the parameter GAIN $= 14.9\text{--}10^{-3}$, the voltages at nodes M and K in the time interval $\Delta t = 19.9$ to 20 ms are to be plotted.

Analysis

- PSpice, Edit Simulation Profile
- Simulation Settings – Fig. 9.10: Analysis,
- Analysis type: Time Domain (Transient)
- Run to time: 20 ms
- Start saving data after: 19.9 ms
- Maximum step size: 0.2 us
- Apply: OK
- PSpice, run

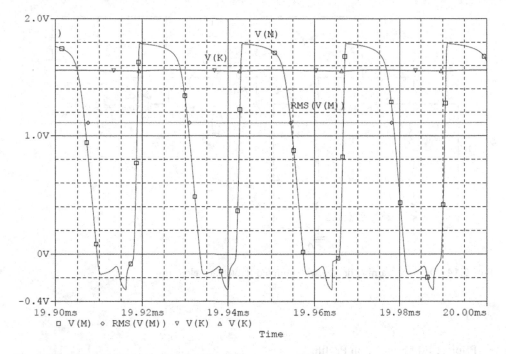

Fig. 9.13   Simulated stresses at nodes M and K

The diagram according to Fig. 9.13 shows the AC voltage and its rms value at node M and the rectified voltage at node K. Due to the voltage drops occurring at both diodes, voltage doubling is not achieved at small amplitudes.

9.3.2   Distance Warning with Inductance Coupling Element

The circuit shown in Fig. 9.14 corresponds to an extension of the receiver in Fig. 9.3. In terms of simulation, the inductance elements are coupled in the equivalent circuit of the transmitter and receiver. This coupling is done with the circuit element K_Linear from the ANALOG library. The LED lights up with the parameter values $k \geq 0.039$.

Task: Circuit Variables with Different Degree of Coupling
For the circuit shown in Fig. 9.14, the following are to be analyzed in the time range $\Delta t = 0$ to 2.5 ms with a variation $k = 0.2$ and 0.4: the LED current and the rms values of the voltages at the transmitter T and receiver R.

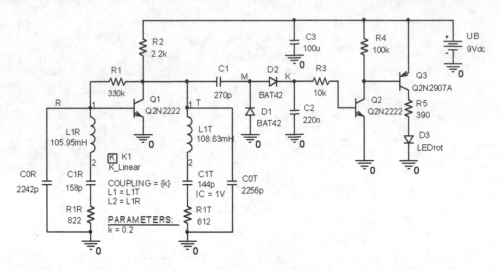

Fig. 9.14 Distance warning device with coupled inductance elements

Analysis

- PSpice, Edit Simulation Profile
- Simulation Settings – Fig. 9.14: Analysis,
- Analysis type: Time Domain (Transient)
- Run to time: 2.5 ms
- Start saving data after: 0
- Maximum step size: 0.2 us
- Apply: OK
- PSpice, run

The analysis shows with Fig. 9.15 that the LED current starts with a time delay at lower coupling, i.e. at greater distance of the two US converters from the reflector wall. In the steady state, the RMS value of the voltage at the transmitter reaches the higher value at lower coupling, see Fig. 9.16. According to Fig. 9.17, the voltage at the receiver increases at stronger coupling.

Task: Stresses at Nodes M and K with Different Couplings
Using the circuit shown in Fig. 9.14, analyze the rms voltage at node M and the rectified voltage at node K for coupling with $k = 0.2$ and 0.4 in the time range $\Delta t = 0$ to 30 ms.

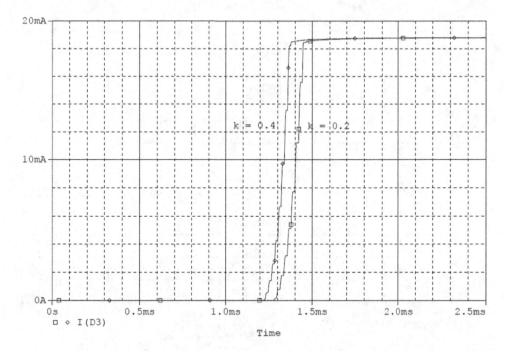

Fig. 9.15 LED current with different coupling

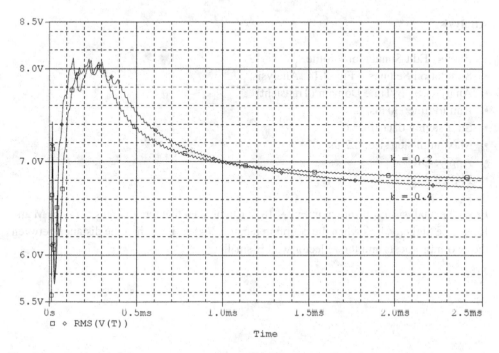

Fig. 9.16 Simulated voltages at the transmitter with different couplings

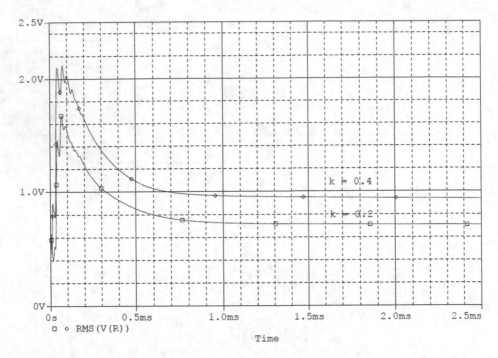

Fig. 9.17 Simulated voltages at the receiver with different coupling

Analysis

- PSpice, Edit Simulation Profile
- Simulation Settings – Fig. 9.14: Analysis,
- Analysis type: Time Domain (Transient)
- Run to time: 30 ms
- Start saving data after: 0
- Maximum step size: 0.2 us
- Apply: OK
- PSpice, run

From the analysis result shown in Fig. 9.18, it can be seen that the voltages at nodes M and K assume higher values when the coupling factor is higher, i.e., when the distance between the two transducers and the reflector wall is smaller.

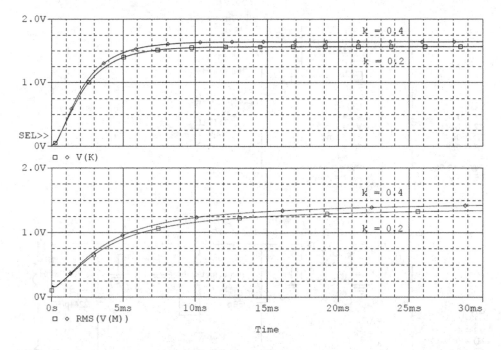

Fig. 9.18 Voltages at the diodes with different coupling

9.4 Transmitting Impulses to the Receiver

In the circuit shown in 9.19, the pulse source U_G generates positive square-wave pulses of 40 kHz at a voltage of 10 V. By multiplicatively combining the source U_G with the source U_M, a packet of 40 square-wave oscillations reaches the transmitter T over the period of one millisecond. The direct transit time between the start of transmission by the transmitter and the onset of the output signal at the receiver is calculated using Eq. (9.2).

$$T_L = \frac{a}{c} \tag{9.2}$$

At a distance $a = 1$ m between transmitter and receiver and with the speed of sound $c = 344$ m/s (at 20 °C in air) one obtains $T_L = 2.907$ ms. This transit time is included in the value for the time T_D of the source U_N. The attenuation of the signal is estimated with the voltage-controlled voltage source E_3 as follows: GAIN = V(R)/V(LC).

For $U_G \approx 7.27$ V_{eff} and $a = 1$ m, the receiver voltage $V(R) \approx 10.72$ mV is obtained via the previously given steps according to [4].

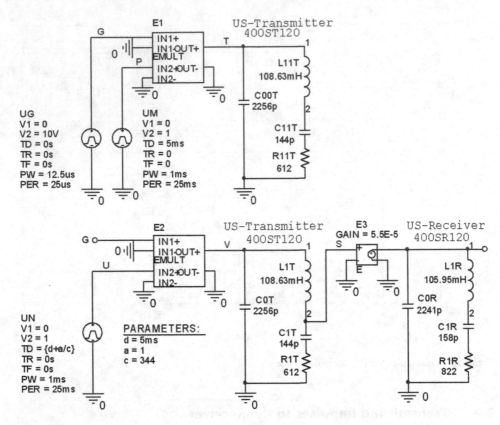

Fig. 9.19 Circuit for transmitting US transmit pulses to the receiver

The simulation provides the voltage V(LC) = 195 V that occurs between L_{11T} and C_{11T} when only the pulse source U_G is connected to the transmitter. Thus GAIN $\approx$ 5.5 E-5 = 5.5–10^{-5}.

Task: Representation of Oscillations
The following vibration packets in the range $\Delta t = 0$ to 40 ms are to be compared:

- the oscillation sections at the circuit nodes T and V.
- the oscillation sections at the circuit nodes T and S.
- the vibration sections at the circuit nodes T and R

Analysis

- PSpice, Edit Simulation Profile
- Simulation Settings – Fig. 9.19: Analysis,
- Analysis type: Time Domain (Transient)

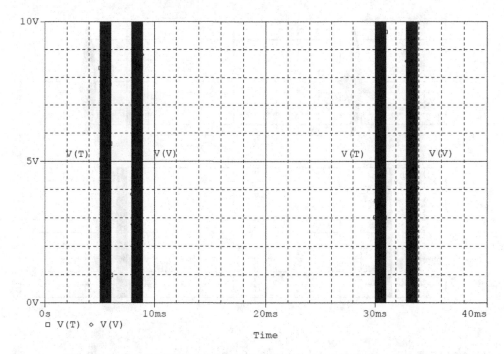

Fig. 9.20 Vibration packets at nodes T and V

- Run to time: 40 ms
- Start saving data after: 0
- Maximum step size: 1 us
- Apply: OK
- PSpice, run

In the representation according to Fig. 9.20, the oscillation packet at node V occurs delayed in comparison to the one at node T by the amount of the transit time $T_L = 2.907$ ms.

Figure 9.21 shows the oscillation interval at node T and the subsequent build-up and decay of large-amplitude sinusoidal signals at node S. The analysis result shown in Fig. 9.22 shows the interval of square-wave oscillations occurring at the transmitter at the circuit input and the delayed and greatly attenuated sinusoidal output signals at the receiver.

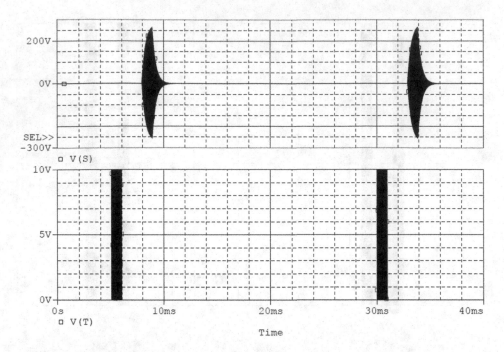

Fig. 9.21 Square-wave transmitted signals and delayed sinusoidal signals at node S

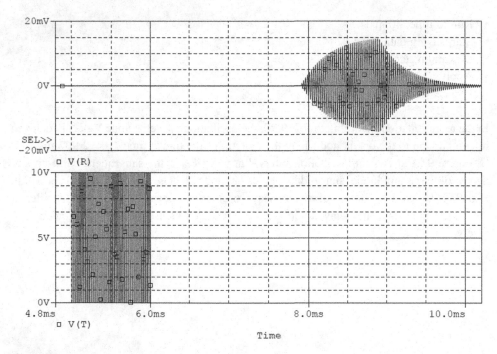

Fig. 9.22 Vibration packets at the transmitter and receiver

References

1. Midas Components Ltd.: Air Ultrasonic Transducers 400ST/R120, Data Sheet (2015)
2. Murata Manufacturing Co: Ultrasonic Sensor. Application Manual (2008)
3. Reichert, A.: Schallversuche. http://chemiephysikskripte.de/schall.htm. Accessed on 08.06.2016 (2014)
4. Pro- Wave Electronics Corp.: Application Note – APO50830 (2015)
5. Stempel, U.E.: Experimente mit Ultraschall. Franzis, München (2012)
6. KEMO- Electronic: Ultraschall- Abstandswarner Bausatz B214 (2014)

Surface Acoustic Wave Devices

10

Surface acoustic wave devices are used as delay lines, filters and resonators up to high frequencies. Furthermore, they serve as sensors for the detection of temperature, pressure or chemical substances. Surface acoustic waves (AOW, SAW) can be generated by an interdigital tranducer (IDT) on a piezoelectric substrate such as quartz, lithium niobate or zinc oxide based on the piezoelectric effect [1–6].

10.1 AOW Delay Lines

AOW delay lines with a quartz substrate and a lithium niobate substrate are considered. It is investigated which different parameter values of the two substrate designs lead to an approximately equal synchronous frequency.

10.1.1 AOW Delay Line with Quartz Substrate

AOW delay lines are used in the frequency range $\Delta f = 20$ MHz to 2 GHz. Their delay times range from $\Delta t_v = 100$ ns to 10 µs and the insertion loss values have an interval $\Delta a_v = 3$ to 35 dB. The insertion loss increases with increasing bandwidth and delay. Because the acoustic propagation velocity v_0 is significantly smaller than the velocity of the electromagnetic waves, quite large delay times t_v can be realized with AOW delay lines for a given transducer spacing l according to Eq. (10.1).

© The Author(s), under exclusive license to Springer Fachmedien Wiesbaden GmbH, part of Springer Nature 2023
P. Baumann, *Selected Sensor Circuits*,
https://doi.org/10.1007/978-3-658-38212-4_10

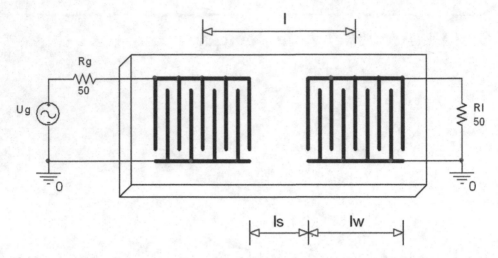

Fig. 10.1 AOW delay line

$$t_v = \frac{l}{v_0} \tag{10.1}$$

Figure 10.1 shows the basic structure of the AOW delay line. The sinusoidal voltage applied to the input causes IDT_1 to generate mechanical waves. At IDT_2 these waves are converted back into an electrical signal via the piezoelectric effect [1–4].

The synchronous frequency f_0 is determined by the propagation velocity v_0 and the wavelength λ according to Eq. (10.2).

$$f_0 = \frac{v_0}{\lambda} \tag{10.2}$$

The number of finger pairs N_p follows with Eq. (10.3) from the required zero point bandwidth B_n. This bandwidth results from the frequency response of the normalized transfer function $H(f)/H(f_0)$ and corresponds to that frequency section on both sides of the synchronous frequency f_0, for which the aforementioned transfer quotient has the value zero, see also Fig. 10.5.

$$N_p = \frac{2}{B_n} \cdot f_0 \tag{10.3}$$

The transducer length l_w follows from the size N_p, the finger spacing a_f and the finger width b_f, see Eq. (10.4) and Fig. 10.2.

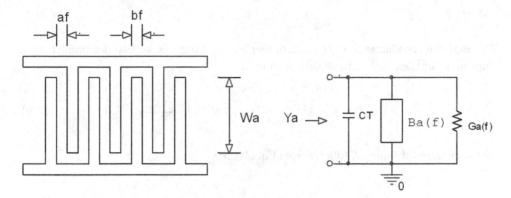

Fig. 10.2 Dimensions of the IDT and the equivalent circuit according to MASON

$$l_w = 2 \cdot N_p \cdot (a_f + b_f) \tag{10.4}$$

The finger width b_f is determined by the wavelength λ according to Eq. (10.5). The finger spacing a_f corresponds to the finger width b_f in the example considered.

$$b_f = \frac{\lambda}{4} \tag{10.5}$$

The two interdigital transducers with their transducer length l_w are separated from each other by the sensitive travel distance l_s. The average distance between these IDTs is described by Eq. (10.6).

$$l = l_w + l_s \tag{10.6}$$

The sensitive length l_s is set as a multiple of the wavelength λ. The height of the finger overlap W_a shown in Fig. 10.2 results from the optimization with Eq. (10.7). This equation and the following calculations and models are based on the explanations of [1, 2].

$$W_a = \frac{1}{R_g} \cdot \frac{1}{2 \cdot f_0 \cdot C_f \cdot N_p} \cdot \frac{4 \cdot k^2 \cdot N_p}{\left(4 \cdot k^2 \cdot N_p\right)^2 + \pi^2} \tag{10.7}$$

C_f denotes the length-related finger capacitance and k represents the coupling factor.

The Mason impulse response model is used to simulate the behaviour of an AOW delay line. As elements it contains the radiation conductance $G_a(f)$, the acoustic susceptance B_a (f) and the total capacitance C_T.

The total capacity C_T is captured by Eq. (10.8).

$$C_T = C_f \cdot W_a \cdot N_p \tag{10.8}$$

The radiation conductance $G_a\,(f)$ according to Eq. (10.9) is largely determined by a sinusoidal function with normalized detuning X.

$$G_a(f) = 8 \cdot k^2 \cdot C_f \cdot W_a \cdot f_0 \cdot N_p{}^2 \cdot \left| \frac{\sin{(X)}^2}{X} \right| \tag{10.9}$$

The normalized detuning X follows from Eq. (10.10).

$$X = N_p \cdot \pi \cdot \frac{(f - f_0)}{f_0} \tag{10.10}$$

The acoustic susceptance $B_a\,(f)$ describes Eq. (10.11).

$$B_a(f) = \frac{G_a(f_0) \cdot \sin{(2 \cdot X)} - 2 \cdot X}{2 \cdot X^2} \tag{10.11}$$

For the single transducer (IDT), the transfer function $H(f)$ is given by Eq. (10.12).

$$H(f) = 2 \cdot k \cdot \sqrt{C_f \cdot f_0} \cdot N_p \cdot \frac{\sin{(X)}}{X} \cdot e^{-j2 \cdot \pi \cdot f \cdot N_p/(2 \cdot f_0)} \tag{10.12}$$

The transfer function $H(f)$ of the AOW delay line according to Eq. (10.13) considers the transfer functions of IDT_1 and IDT_2. Via the length l according to Eq. (10.6), the delay line l_s is included. This length l_s can be fixed with some multiples of the wavelength λ.

$$H(f) = H_1(f) \cdot H_2(f) \cdot e^{-j2 \cdot \pi \cdot f \cdot l/v_0} \tag{10.13}$$

The magnitude of the normalized transfer function for the two identical IDTs follows from Eq. (10.14).

$$\left| \frac{H(f)}{H(f_0)} \right| = \left| \left(\frac{\sin{(X)}}{X} \right)^2 \right| \tag{10.14}$$

The insertion loss a_v (Insertion loss IL) is determined by the acoustic admittance $Y_a = G_a + jB_a$ as well as determined by the total capacitance C_T and the load resistance R_g, see Eq. (10.15).

$$a_v = -10 \cdot \log \left[\frac{2 \cdot G_a(f) \cdot R_g}{\left(1 + G_a(f) \cdot R_g\right)^2 + \left[R_g \cdot (2 \cdot \pi \cdot f \cdot C_T + B_a(f))\right]^2} \right] \qquad (10.15)$$

Task: Calculation of Parameters
Given are the material parameters of an AOW delay line (quartz, ST-cut) with $v_0 = 3158$ m/s, $k = 0.04$ and $C_f = 50.70$ pF/m [1]. The wavelength following from the finger dimensions is $\lambda = 48$ µm, and the zero-point bandwidth is given as $B_n = 1.5$ MHz. Furthermore, the sensitive length is selected with $l_s = 50 \lambda$. The synchronous frequency, the number of finger pairs, the transducer length, the amount of finger overlap, the total capacitance, and the delay time are to be calculated.

Result: $f_0 = 65{,}792$ MHz, $N_p \approx 88$, $l_w = 4.224$ mm, $W_a = 1.883$ mm, $C_T = 8.34$ pF, $t_v = 1.41$ µs.

Task: Analysis of Frequency-Dependent Variables
In the frequency range $\Delta f = 62.5$ to 69 MHz, the following characteristics of the AOW delay line with quartz substrate are to be analyzed according to Fig. 10.3:
The radiation conductance $G_a(f)$, the acoustic susceptance $B_a(f)$, the magnitude of the normalized transfer function $H_n = H(f)/H(f_0)$ and the transmission loss a_v.

Solution
Figure 10.3 shows for each parameter a circuit with a voltage source, where the original value of 1 Vdc is replaced by the designation of the parameter placed in curly brackets. The respective equation is entered under PARAMETERS from the library SPECIAL. A DC sweep analysis is to be performed. This analysis is only used for the formal processing of the previously specified equations. For this purpose, the frequency f is introduced as a global parameter.

Analysis

- PSpice, Simulation profile
- Simulation Settings – Fig. 10.3: Analysis
- Analysis type: DC Sweep
- Options: Primary Sweep
- Sweep variable: Global Parameter
- Parameter Name: f
- Sweep type: Linear
- Start value: 62.5 Meg
- End value: 69 Meg
- Increment: 10 k

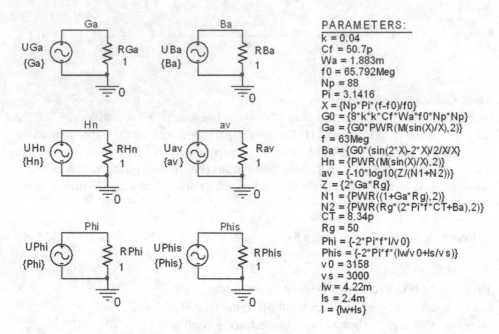

PARAMETERS:
k = 0.04
Cf = 50.7p
Wa = 1.883m
f0 = 65.792Meg
Np = 88
Pi = 3.1416
X = {Np*Pi*(f-f0)/f0}
G0 = {8*k*k*Cf*Wa*f0*Np*Np}
Ga = {G0*PWR(M(sin(X)/X),2)}
f = 63Meg
Ba = {G0*(sin(2*X)-2*X)/2/X/X}
Hn = {PWR(M(sin(X)/X),2)}
av = {-10*log10(Z/(N1+N2))}
Z = {2*Ga*Rg}
N1 = {PWR((1+Ga*Rg),2)}
N2 = {PWR(Rg*(2*Pi*f*CT+Ba),2)}
CT = 8.34p
Rg = 50
Phi = {-2*Pi*f*l/v0}
Phis = {-2*Pi*f*(lw/v0+ls/vs)}
v0 = 3158
vs = 3000
lw = 4.22m
ls = 2.4m
l = {lw+ls}

Fig. 10.3 Characteristics of the AOW delay line with quartz substrate

- Apply: OK
- PSpice, run

The execution of the analysis with the frequency as a global parameter according to Fig. 10.4 shows the frequency response of the admittance $Y_a(f)$. The radiation conductance reaches the value $G_a(f_0) = 622$ µS at the synchronous frequency f_0 and at this frequency the acoustic susceptance takes the value $B_a(f_0) = 0$.

Figure 10.5 shows the frequency response of the magnitude of the normalized transfer function. At $H_n(f) = 0$, one can take the specified value of the zero-sequence bandwidth $B_n = 1.5$ MHz. The minimum value of the insertion loss appears at the synchronous frequency with $a_v(f_0) = 12.44$ dB, see Fig. 10.6.

10.1.2 AOW Delay Line with Lithium Niobate Substrate

In the following, we will consider which characteristics of the AOW delay line with lithium niobate substrate (LiNbO$_3$, YZ) must be changed in order to achieve a synchronous frequency that is approximately the same as that of the previously discussed example of the AOW delay line with quartz substrate (ST section).

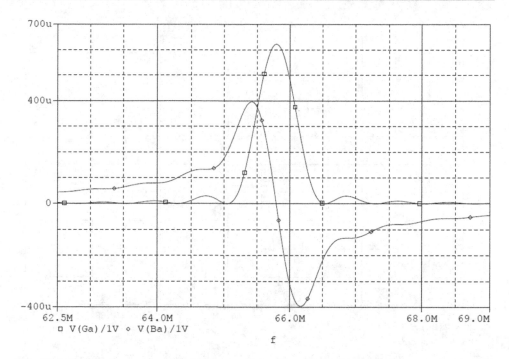

Fig. 10.4 Frequency response of radiation conductance and acoustic susceptance

Task: Calculation of Parameters

With $v_0 = 3488$ m/s, the synchronous frequency $f_0 = 67,076$ MHz follows from Eq. (10.2) for the wavelength $\lambda = 52$ µm. (For the quartz substrate, $f_0 = 65.79$ MHz). The finger width and finger spacing follow from Eq. (10.5) with $a_f = b_f = 13$ µm. Given is a relatively large zero width with $B_n = 4.5$ MHz. This results in a relatively small number of finger pairs with $N_p = 30$, thus keeping the total capacitance C_T small. The height of the finger overlap is set as $W_a = 45-\lambda = 2.34$ µm, because the optimization Gl. Equation (10.7) gives too low values for the LiNbO$_3$ substrate. The sensitive length is set as $l_s = 60-\lambda = 3.12$ mm. From Eq. (10.4), the transducer length emerges as $l_w = 1.56$ mm and the delay time is $t_v = 1.34$ µs.

Furthermore, the piezoelectric characteristic $k^2 = 4.1\%$ [5] and the high value of the length-related finger capacitance $C_f = 460$ pF/m apply to YZ lithium niobate.

To be calculated are the total capacitance C_T and the value of the radiation conductance $G_a (f)$ for detuning $X = 0$.

Result: $C_T = 32.29$ pF, $G_a (f_0) = G_0 = 21.32$ mS.

Task: Analysis of Frequency-Dependent Parameters

Using the circuits shown in Fig. 10.7, the following are to be analyzed in the frequency range $\Delta f = 55$ to 80 MHz: the radiation conductance $G_a (f)$, the acoustic susceptance B_a

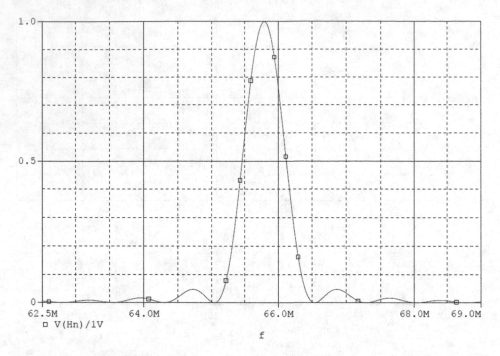

Fig. 10.5 Frequency response of the normalized transfer function

(f), the magnitude of the normalized transfer function $H_n = H(f)/H(f_0)$ and the transmission loss a_v.

Analysis

- PSpice, Simulation profile
- Simulation Settings – Fig. 10.7: Analysis
- Analysis type: DC Sweep
- Options: Primary Sweep
- Sweep variable: Global Parameter
- Parameter Name: f
- Sweep type: Linear
- Start value: 55 Meg
- End value: 80 Meg
- Increment: 10 k
- Apply: OK
- PSpice, run

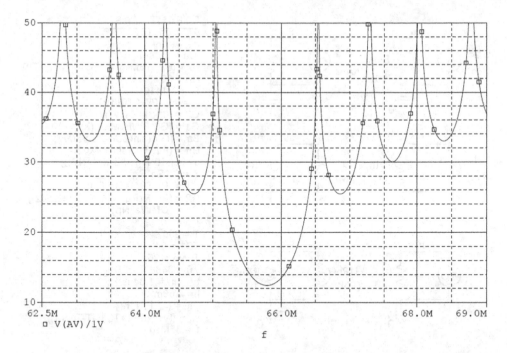

Fig. 10.6 Frequency response of insertion loss

In Fig. 10.8 it can be seen that the values of the acoustic admittance Y_a (f) $= G_a$ (f) $+ jB_a$ (f) are higher for the delay line with the YZ lithium niobate substrate than for that with quartz substrate. This is due to the higher values of the piezoelectric coefficient k and the length-related finger capacitance C_f. One can see the previously calculated maximum value $Ga(f_0) = G_0 = 21.32$ mS. It is $R_0 = 1/G_0 = 46.9$ Ω.

Figure 10.9 shows the specified zero point bandwidth $B_n = 4.5$ MHz.
The insertion loss shows the low value of about three decibels, see Fig. 10.10.

10.1.3 Sensor Applications

10.1.3.1 AOW Temperature Sensor with Lithium Niobate Substrate
With the high temperature coefficient $TC_1 = 94$ ppm/K, the AOW delay line with YZ lithium niobate substrate is suitable as a temperature sensor. The temperature coefficient TC_1 of the delay time t_v describes Eq. (10.16).

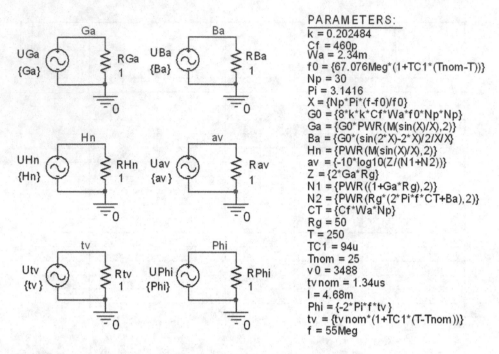

PARAMETERS:
k = 0.202484
Cf = 460p
Wa = 2.34m
f0 = {67.076Meg*(1+TC1*(Tnom-T))}
Np = 30
Pi = 3.1416
X = {Np*Pi*(f-f0)/f0}
G0 = {8*k*k*Cf*Wa*f0*Np*Np}
Ga = {G0*PWR(M(sin(X)/X),2)}
Ba = {G0*(sin(2*X)-2*X)/2/X/X}
Hn = {PWR(M(sin(X)/X),2)}
av = {-10*log10(Z/(N1+N2))}
Z = {2*Ga*Rg}
N1 = {PWR((1+Ga*Rg),2)}
N2 = {PWR(Rg*(2*Pi*f*CT+Ba),2)}
CT = {Cf*Wa*Np}
Rg = 50
T = 250
TC1 = 94u
Tnom = 25
v0 = 3488
tvnom = 1.34us
l = 4.68m
Phi = {-2*Pi*f*tv}
tv = {tvnom*(1+TC1*(T-Tnom))}
f = 55Meg

Fig. 10.7 Characteristics of the delay line with lithium niobate substrate

$$TC_1 = \frac{1}{t_v} \cdot \frac{\Delta t_v}{\Delta T} = -\frac{1}{f} \cdot \frac{\Delta f}{\Delta T} \qquad (10.16)$$

The phase change $\Delta\varphi$ in units of radians follows from Eq. (10.17).

$$\Delta\varphi = -2 \cdot \pi \cdot f \cdot \Delta t_v \qquad (10.17)$$

The conversion of the phase change from radians to degrees is given by Eq. (10.18).

$$\varphi = -2 \cdot \pi \cdot f \cdot \Delta t_v \cdot \frac{180}{\pi} = -360° \cdot f \cdot \Delta t_v \qquad (10.18)$$

Task: Temperature Dependence of the Delay Time

Based on the circuit shown in Fig. 10.7, show the temperature dependence of the delay time at frequency $f = f_0 = 67,076$ MHz in the range $\Delta T = 25$ to 250 °C.

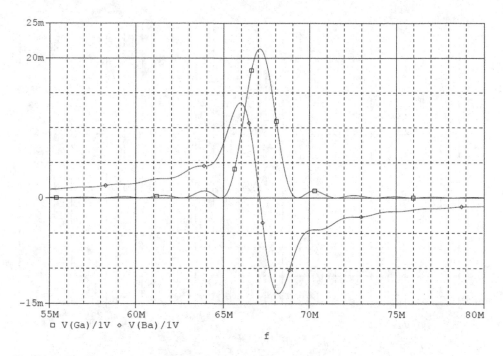

Fig. 10.8 Frequency response of radiation conductance and acoustic susceptance

Analysis

- PSpice, Simulation profile
- Simulation Settings – Fig. 10.7: Analysis
- Analysis type: DC Sweep
- Options: Primary Sweep
- Sweep variable: Global Parameter
- Parameter Name: f
- Sweep type: Logarithmic, Decade
- Start value: 67.076 Meg
- End value: 67.076 Mcg
- Points/Decade: 1
- Options: Parametric Sweep
- Sweep variable: Global Parameter
- Parameter Name: T
- Sweep type: linear
- Start value: 25
- End value: 250
- Increment: 0.1
- Apply: OK
- PSpice, run

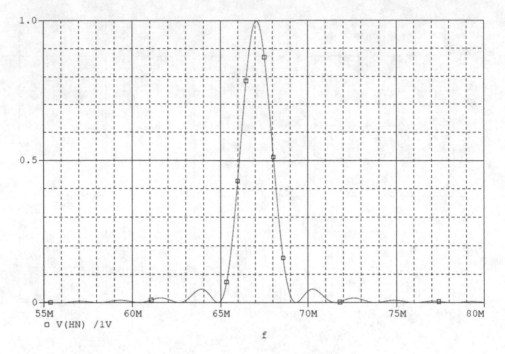

Fig. 10.9 Frequency response of the normalized transfer function

Figure 10.11 shows the linear increase of the delay time when the temperature is increased. In the example, the delay time increases at $f = f_0$ from $t_v = 1.34$ μs at 25 °C to 1.3683 μs at 250 °C.

Task: Temperature Dependence of the Insertion Loss
Using the circuit shown in Fig. 10.7, analyze the insertion loss a_v for the temperature values $T = 25$ and 250 °C.

Analysis

- PSpice, Simulation profile
- Simulation Settings – Fig. 10.7: Analysis
- Analysis type: DC Sweep
- Options: Primary Sweep
- Sweep variable: Global Parameter
- Parameter Name: f
- Sweep type: Linear
- Start value: 55 Meg
- End value: 80 Meg

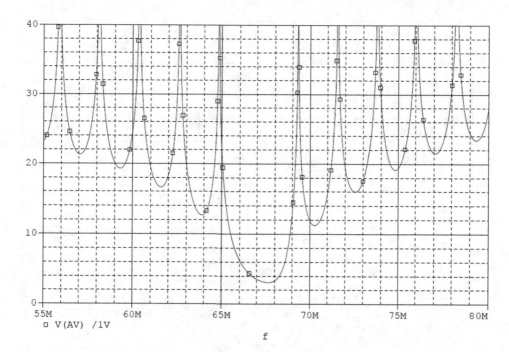

Fig. 10.10 Frequency response of insertion loss

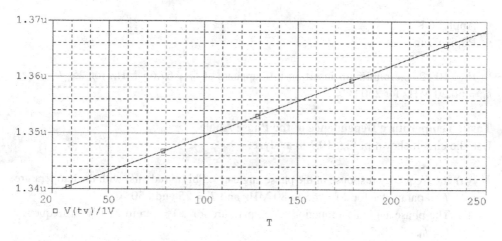

Fig. 10.11 Temperature dependence of the delay time

- Increment: 10 k
- Options: Parametric Sweep
- Sweep variable: Global Parameter
- Parameter Name: T
- Sweep type: Value list: 25250

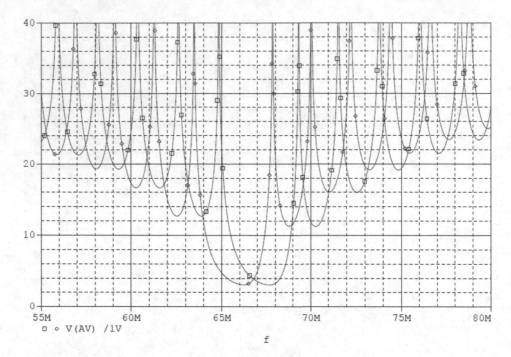

Fig. 10.12 Frequency response of insertion loss at 25 and 250 °C

- Apply: OK
- PSpice, run

In Fig. 10.12, the minimum insertion loss appears at $f_0 = 67.65$ MHz for 25 °C and at $f_0 = 66.20$ MHz for 250 °C.

Task: Temperature Dependence of the Phase Angle
Use the circuit shown in Fig. 10.7 to analyze:

  (a) the frequency response of the phase angle $\varphi =$ Phi (in radians) with the temperature T as parameter for $\Delta f = 55$ to 80 MHz and $T = 25$ and 250 °C.
  (b) The phase angle as a function of temperature for $\Delta T = 25$ to 250 °C at frequency $f = f_0$.

The analysis for (a) is to be carried out analogously to the task for insertion loss.
    The analysis for (b) corresponds to that for the delay time.
    The analysis result according to Fig. 10.13 gives, for $f = f_0$, the phase change $\Delta \varphi = -11.94$ rad at $\Delta T = 225$ °C.
    The temperature dependence of the phase angle in radians at the synchronous frequency f_0 can be seen from the analysis result shown in Fig. 10.14.

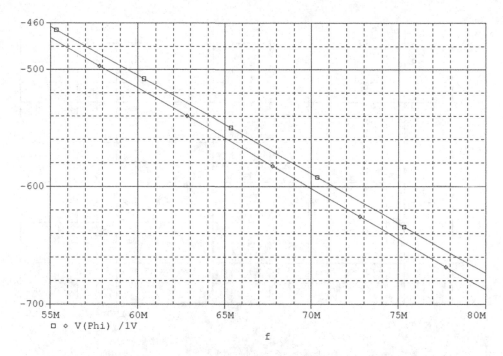

Fig. 10.13 Frequency response of the phase angle in radians at 25 and 250 °C

Task: Temperature Dependence of the Admittance
Using the circuits in Fig. 10.7, analyze the frequency response of the acoustic admittance Y_a (f) = $G_a(f) + jB_a(f)$ for the range $\Delta f = 55$ to 80 MHz with temperature as a parameter for $T = 25$ and 250 °C. From the result, plot the frequency locus curves for both temperatures.

Analysis
The analysis is the same as for insertion loss.
   The analysis result according to Fig. 10.15 shows:

- At $T = 250$ °C, $f_{0T} = 65{,}655$ MHz is smaller than $f_0 = 67{,}076$ MHz at $T = 25$ °C.
- From $f_{0T} < f_0$ follows a lower maxlmum of $G_a(f)$ for the higher temperature.
- The acoustic susceptance $B_a(f)$ follows the temperature trend of $G_a(f)$ by Eq. (10.11).

The frequency locus curves of the acoustic admittance show only minor differences. However, in Fig. 10.16 for $T = 25$ °C and disappearing imaginary part the frequency f_0 appears, while in this case according to Fig. 10.17 at $T = 250$ °C the smaller frequency f_{0T} is present. The maxima and mlnima of the susceptance $B_a(f)$, which lie at different frequencies, should also be noted. The upper half of the respective locus curve describes the capacitive and the lower half the inductive behaviour of the acoustic delay line.

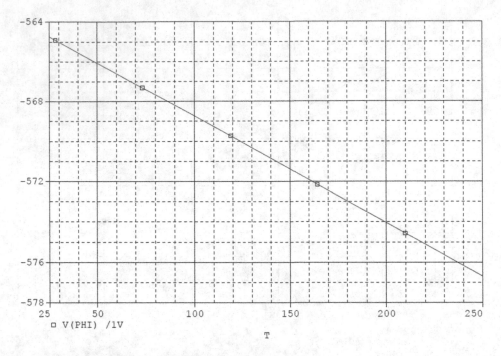

Fig. 10.14 Temperature dependence of the phase angle in Radiant

10.1.3.2 Sensor with Quartz Substrate for the Detection of Chemical Substances

AOW delay lines with a substrate of quartz (ST section) are suitable with the sensitive membrane as a sensor for the detection of chemical substances such as exhaust gases, vapours, and deposits of layers. On the sensitive section, the velocity of the surface wave v_s is smaller than the propagation velocity v_0 in many applications. Equation (10.19) is used to obtain the sensitively influenced phase angle (in radians).

$$\varphi_s = -2 \cdot Pi \cdot f \cdot \left(\frac{l_w}{v_0} + \frac{l_s}{v_s} \right) \tag{10.19}$$

Task: Detection of a Chemical Coating

With the circuits shown in Fig. 10.3, in the frequency range $\Delta f = 62$ to 69 MHz for the delay line are to be analyzed:

• the phase angle φ (in radians) for the uncoated retardation distance with $v_0 = 3158$ m/s
• the phase angle φ_s for a coated retardation section with $v_s = 3000$ m/s

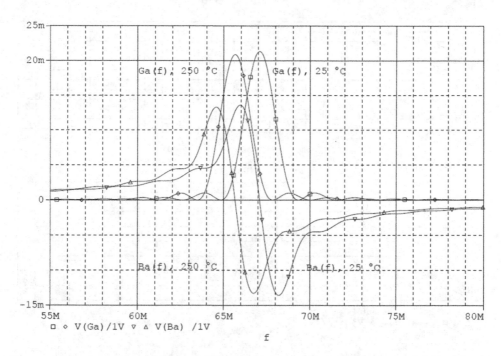

Fig. 10.15 Frequency response of conductance and susceptance at 25 and 250 °C

Analysis

- PSpice, Simulation profile
- Simulation Settings – Fig. 10.3: Analysis
- Analysis type: DC Sweep
- Options: Primary Sweep
- Sweep variable: Global Parameter
- Parameter Name: f
- Sweep type: Linear
- Start value: 62 Meg
- End value: 69 Meg
- Increment: 10 k
- Apply: OK
- PSpice, run

As a result of the analysis, Fig. 10.18 shows the respective linear decrease of the phase angle when the frequency is increased.

For $f = f_0 = 65,792$ MHz holds:

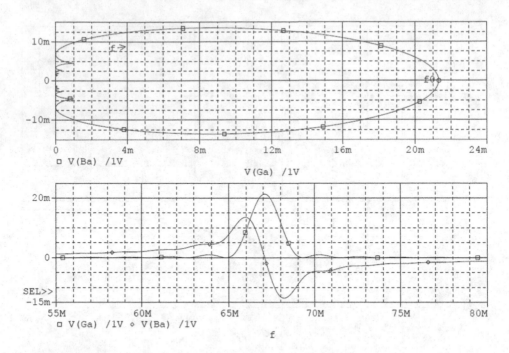

Fig. 10.16 Acoustic admittance and frequency locus curve at 25 °C

- $T = 25\,°\text{C}$ is $\varphi = -866.59\,\text{rad} = -49{,}652°$.
- $T = 250\,°\text{C}$, one obtains $\varphi = -883.14\,\text{rad} = -50{,}600°$.

The lower propagation speed of the sensitive layer leads to an evaluable difference in the phase angles.

10.2   Surface Acoustic Wave Resonators

Surface acoustic wave resonators can be used as frequency stabilizing devices in a range from 200 to 1000 MHz. The one-port resonator is a typical component of Colpitts oscillators.

The frequency response of the two-port resonator corresponds to that of a band-pass filter with a very narrow bandwidth and low insertion loss.

(Insertion Loss, *IL*). This component is used both as a narrow bandwidth filter and as an element of oscillators.

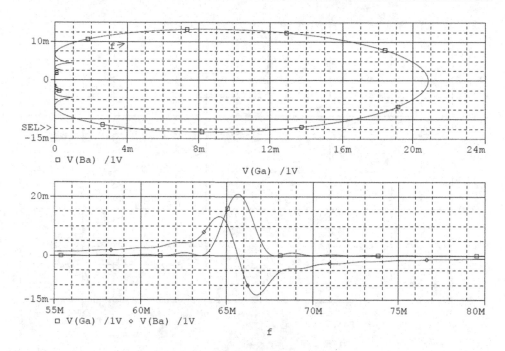

Fig. 10.17 Acoustic admittance and frequency locus at 250 °C

10.2.1 One-Port Surface Wave Resonator

Figure 10.19 shows the structural design of the one-port AOW resonator [7]. When an electrical signal is applied to the interdigital transducer, the acoustic waves emitted by it are reflected back by the reflectors on both sides, thus generating a standing wave. The frequency response, bandwidth and attenuation depend, among other things, on the substrate material used (quartz, lithium niobate, lithium tantalate) as well as on the number and spacing of the finger bars of the interdigital transducer.

The starting point for the following analyses is the SR315 single-port resonator from VANLONG [8]. The nominal frequency f_n given in the data sheet is measured by the manufacturer as the frequency at which the minimum insertion loss s_{21} (Insertion Loss IL) occurs in the 50 Ohm test system used. Furthermore, it is pointed out that the RLC model only approximates the resonator behaviour near the resonant frequency.

Data Sheet

For the single-port AOW resonator SR315 are given by Vanlong [8]:

- $f_n = 315.00$ MHz $+/-$ 75 kHz, $IL = 1.5$ dB < 2.0 dB, $Q_U = 12{,}500$, $Q_L = 2000$
- $R_m = 19\,\Omega$, $L_m = 120.3114\,\mu$H, $C_m = 2.124$ fF, $C_0 = 2.6$ pF (2.3 to 2.9 pF).

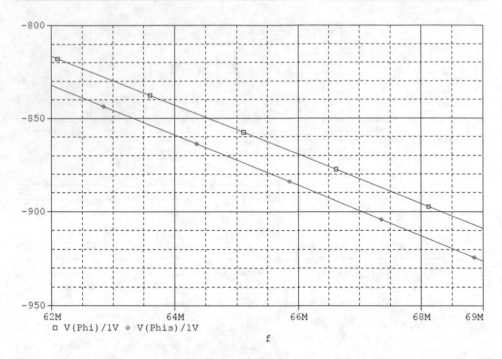

Fig. 10.18 Phase angle without and with sensitive coating

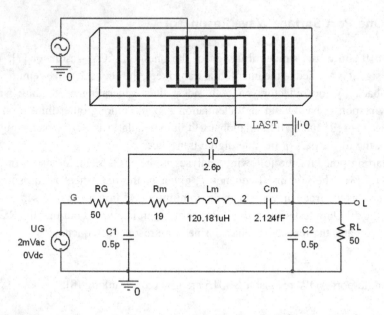

Fig. 10.19 Design and test circuit of the one-port AOW resonator

Using the data sheet element values, the circuit shown in Fig. 10.19 yields the insertion loss with $s_{21} = IL = -1.4415$ dB at $f = 314,829$ MHz.

To obtain the minimum insertion loss at $f_c = \textbf{315,000 MHz}$, L_m and/or C_m would have to be reduced. The circuit shown in Fig. 10.19 achieves this goal as a variant for $L_m = \textbf{120.1810 μH}$ (instead of 120.3114 μH), see Fig. 10.19.

Task: Frequency Dependence of the Insertion Loss
Given the circuit shown in Fig. 10.19, for the frequency range $\Delta f = 314$ to 316 MHz, analyze the forward transfer factor s_{21} by magnitude and phase.

Analysis

- PSpice, Simulation profile
- Simulation Settings – Fig. 10.19: Analysis
- Analysis type: AC Sweep/Noise
- AC Sweep type; Linear
- Start Frequency: 116 Meg
- End Frequency: 118 Meg
- Points/Decade: 5 k
- Apply: OK
- PSpice, run

In Fig. 10.20, at $f = 315,000$ MHz, the minimum value of the attenuation appears with $|s_{21}| = -1.4414$ dB and the phase angle $\varphi_{21} = 4.7301°$.

When used in the Colpitts oscillator, the input of the one-port AOW resonator is connected to the base of the RF transistor and the output is connected to ground [9]. This gives the circuit shown in Fig. 10.21.

Some manufacturers [9, 10] define the center frequency f_c as the frequency at which the real part Re (y_{11}) of the input conductance y_{11} has its maximum. Using all the element values as in the circuit shown in Fig. 10.20, the resonance is obtained at $f = 315,003$ MHz. Reducing the inductance to $L_m = \textbf{120.1835 μH}$ leads to the center frequency $f_c = \textbf{315,000 MHz}$. The proof is provided with the following analysis.

Task: Frequency Dependence of the Input Admittance
For the circuit shown in Fig. 10.21, the frequency dependence of the input admittance by real and imaginary parts in the range $\Delta f = 314.7$ to 315.3 MHz shall be analyzed and evaluated.

Analysis

- PSpice, Simulation profile
- Simulation Settings – Fig. 10.21: Analysis

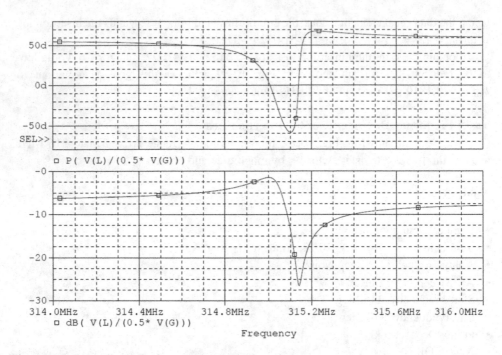

Fig. 10.20 Forward transfer factor with magnitude and phase

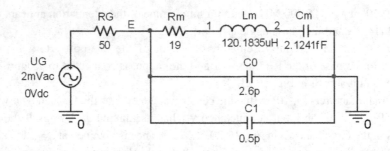

Fig. 10.21 Equivalent circuit of the resonator with short-circuited output

- Analysis type: AC Sweep/Noise
- AC Sweep type; Linear
- Start Frequency: 314.7 Meg
- End Frequency: 315.3 Meg
- Points/Decade: 5 k
- Apply: OK
- PSpice, run

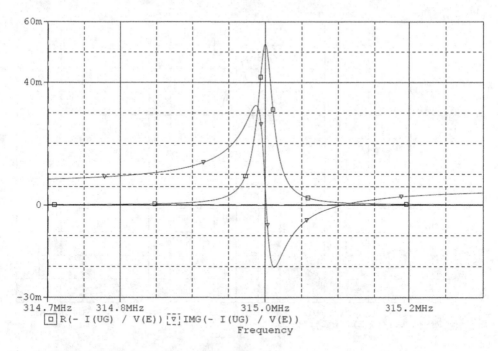

Fig. 10.22 Real part and imaginary part of the input admittance

The analysis result shown in Fig. 10.22 gives the real part Re (Y) = 52.631 mS and the imaginary part Im (Y) = 5.9852 mS at the center frequency f_c = **315,000 MHz**. Furthermore, 1/Re (Y) = R_m = 19.00 Ω and from the imaginary part follows the capacitance C = 3.024 pF.

From the frequency dependence of the real part Re(Y), the unloaded Q_u can be determined according to Eq. (10.20) [11]. It holds

$$Q_u = \frac{f_c}{f_2 - f_1} \tag{10.20}$$

Here f_c is the center frequency and f_1 and f_2 are those frequencies resulting from the intersection of the Re(Y) frequency dependence with the line for Re(Y_{max})/2, see Fig. 10.23. The analysis is to be performed as for Fig. 10.22. Using the analysis results f_c = 315,000 MHz, f_1 = 314,987 MHz, and f_2 = 315,012 MHz, it follows that Δf = 0.025 MHz and hence Q_u = 12,600. This Q can also be determined from the RLC parameters using the values of Fig. 10.21 according to Eq. (10.21).

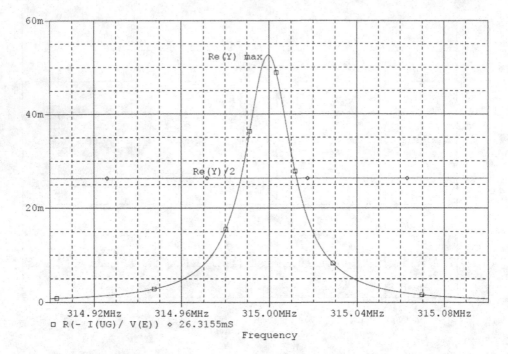

Fig. 10.23 Determination of the unloaded quality of the one-port resonator

$$Q_u = \frac{1}{R_m} \cdot \sqrt{\frac{L_m}{C_m}} \tag{10.21}$$

You calculate $Q_u = 12{,}519$.

10.2.2 Colpitts Oscillator

Figure 10.24 shows the application of the one-port AOW resonator SR315 in a Colpitts oscillator circuit [12, 13].

The oscillation frequency of the Colpitts oscillator is calculated using Eq. (10.22).

$$f_0 = \frac{1}{2 \cdot \pi \cdot \sqrt{L \cdot C_1 \cdot C_2/(C_1 + C_2)}} \tag{10.22}$$

In the considered circuit, the inductance L is provided by the resonator, whose impedance shows inductive behavior in the range between the series and parallel resonance [12]. Twenty-five SPICE model parameters of the Infineon BFR 92P UHF bipolar transistor [14] were input to a transistor QbreakN called from the BREAK library.

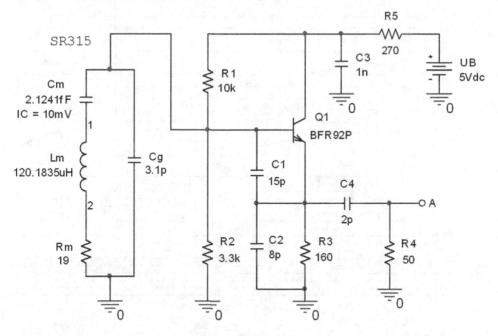

Fig. 10.24 Colpitt's oscillator with one-port AOW resonator

Task: Analysis of the Output Oscillations
For the circuit shown in Fig. 10.24, analyze the frequency of the output oscillations.

Analysis

- PSpice, Edit Simulation Profile
- Simulation Settings- Fig. 10.24: Analysis
- Analysis type: Time Domain (Transient)
- Run to time: 200 ns
- Start saving data after: 0 s
- Maximum step size: 0.005 ns
- Apply: OK
- PSpice run

As an analysis result, Fig. 10.25 shows the oscillations at the oscillator output at the center frequency $f_c = 315,000$ MHz.

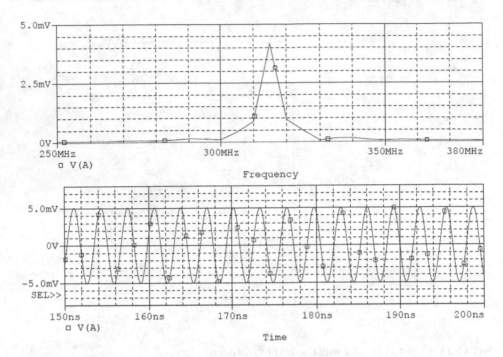

Fig. 10.25 Oscillations at the output of the Colpitts oscillator

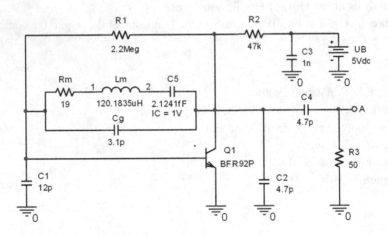

Fig. 10.26 Pierce oscillator with one-port AOW resonator

10.2.3 Pierce Oscillator

Figure 10.26 shows the circuit of a Pierce oscillator [15, 16]. The UHF amplifier transistor BFR92P causes a phase rotation of $-180°$. A further phase rotation of $-180°$ is provided

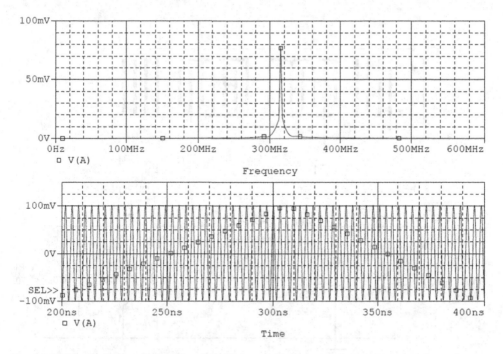

Fig. 10.27 Oscillations at the output of the Pierce oscillator

by the resonator in conjunction with the capacitors C_1 and C_2, thus satisfying the necessary phase condition for oscillation generation. The amplitude condition according to Barkhausen is that the magnitude of the product of the feedback factor k and the amplification v becomes equal to one: $k \cdot v = 1$. The oscillation frequency according to Eq. (10.22) follows from the relation according to Eq. (10.23). Here the magnitude L corresponds to the inductive effect of the resonator in the region between the series and parallel resonance.

$$-\frac{1}{\omega_0 \cdot C_1} - \frac{1}{\omega_0 \cdot C_2} + \omega_0 \cdot L = 0 \qquad (10.23)$$

Task: Analysis of the Oscillation Frequency
For the circuit shown in Fig. 10.26, the oscillations at the circuit output are to be represented. The level of the oscillation frequency is to be determined via Fourier analysis (Fig. 10.27).

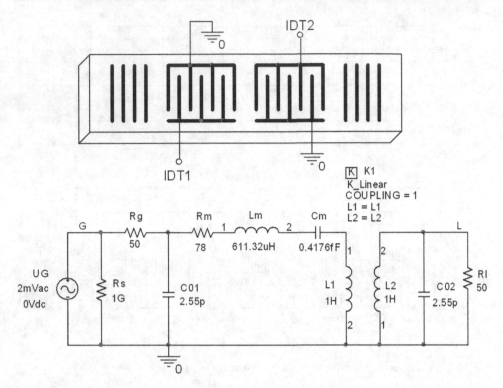

Fig. 10.28 Two-port OFW resonator with structure and equivalent circuit

Analysis

- PSpice, Edit Simulation Profile
- Simulation Settings- Fig. 10.26: Analysis
- Analysis type: Time Domain (Transient)
- Run to time: 400 ns
- Start saving data after: 200 ns
- Maximum step size: 0.005 ns
- Apply: OK
- PSpice run

10.2.4 Two-Port Surface Acoustic Wave Resonator

Figure 10.28 shows the structure of the two-port AOW transducer. On the piezoelectric substrate there are two coupled interdigital transducers and the outer reflectors.

The entered element values belong to the resonator type SQ315 from Vanlong [8]. The value for the inductance was changed compared to the data sheet of $L_m = 611.8694$ µH reduced to 611.32 µH to obtain the minimum insertion loss at $f = 315{,}000$ MHz.

Data Sheet
AOW Resonator SQ315 from Vanlong [8].

$f_n = 315.00$ MHz +/− 75 kHz, $IL = 5.0 < 7.0$ dB, $Q_u = 15{,}520$, $Q_L = 6800$.
$R_m = 78 < 124$ Ω, $L_m = 611.8694$ µH, $C_m = 0.4178$ fF, $C_0 = 2.55$ (2.25–2.85) pF.

The data sheet specifications lead to $|s_{21}| = -5.2491$ dB at $f = 314{,}859$ MHz.

Task: Frequency Dependence of Parameters
With the circuit shown in Fig. 10.28, the frequency range from 314 to 316 MHz can be analysed:

- The forward transmission factor s_{21} according to magnitude and phase
- The input reflection factor s_{11} according to magnitude and phase

Analysis
- PSpice, Simulation profile
- Simulation Settings – Fig. 10.28: Analysis
- Analysis type: AC Sweep/Noise
- AC Sweep type; Linear
- Start Frequency: 314.7 Meg
- End Frequency: 315.3 Meg
- Points/Decade: 5 k
- Apply: OK
- PSpice, run

Using Fig. 10.29, the analysis of the frequency dependence of the forward transfer factor yields the values: $|s_{21}| = -5.2475$ dB and $\varphi_{21} = -208, 367°$ at $f_n = 315{,}000$ MHz.

The frequency dependence of the input reflection coefficient is shown in Fig. 10.30.

At $f = 314{,}996$ MHz, the following are obtained: $|s_{11}| = -6.1381$ dB and $\varphi_{11} = -6.7113°$.

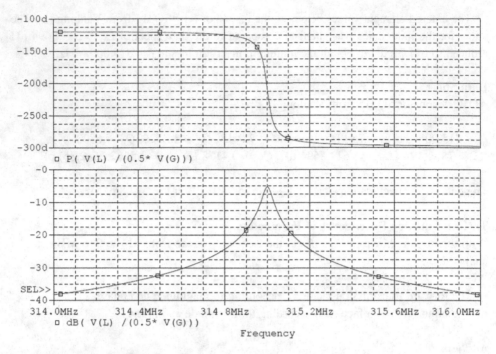

Fig. 10.29 Forward transfer factor with magnitude and phase

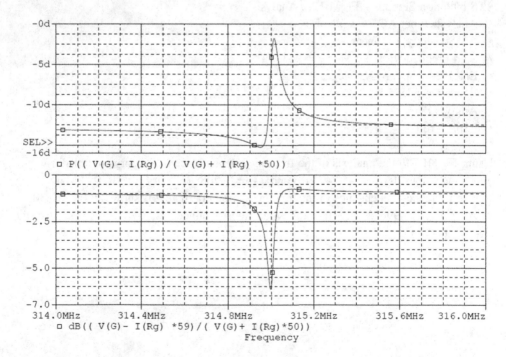

Fig. 10.30 Input reflection factor with magnitude and phase

References

1. Wilson, W.C., Atkinson, G.M.: Modeling Sensors on Aerospace Vehicles with Mathcad (NASA-Center Document ID: 20070016024, May 9, 2007). PTC Express (2007)
2. Wilson, W. C., Atkinson, G. M.: Mixed Modeling of a SAW Delay Line using VHDL. NASA gov.search (2018)
3. Bhattacharyya, A.B.: SAW Device Modelling for SPICE Simulation. GSA India, IIT Delhi (2012)
4. Malik, A.F.: Acoustic Wavelength Effects on the Propagation of SAW on Piezo Crystal. RSM Proc, Langkawi (2013)
5. DEGA DPG Workshop Physikalische Akustik, Bad Honnef (2003)
6. Reindl, L., et al.: Berührungslose Messung der Temperatur mit passiven OFW-Sensoren. VDI-Berichte Nr. 1379 (1998)
7. RF Wireless World: Port SAW Resonator vs 2 Port Saw Resonator (2007)
8. Vanlong Technology Co. Ltd.: Datenblätter zu den SAW-Resonatoren SQ315 und SR315 (2007)
9. EPCOS AG Application Note 25. Design Guide fort he SAW Oscillator. München (2014)
10. SAW Components Dresden: Datenblatt zu SR315B. Dresden (2013)
11. Namdeo, A.K.: Extraction of Electrical Circuit of One Port SAW Resonator Using FEM Based Simulation Proc. of the COMSOL Conference. Pune (2015)
12. Murata Manufacuring Co. Ltd.: Application and Data of SAW Resonator, P36E.pdf02.8.5, (2002)
13. Eisherbini, M.M., et al.: Design and simulation for UHF oscillator using SAWR with different schematics. Indones. J. Electr. Eng. Comput. Sci. 1(2), 495–501 (2016)
14. Infineon: Data Book, Part 2: SPICE-Parameter des Transistors BFR92P, München (2000)
15. Parker, E.T., Montress, G.K.: Precision Surface. Acoustic-Wave (SAW) Oscillators. IEEE Trans. Ultrason. Ferroelectr. Freq. Control. 15(5), 342–349 (1988)
16. Neubig, B., Briese, W.: Das große Quarzkochbuch. Franzis, Feldkirchen (1997)

Index

© The Author(s), under exclusive license to Springer Fachmedien Wiesbaden GmbH, part of Springer Nature 2023
P. Baumann, *Selected Sensor Circuits*,
https://doi.org/10.1007/978-3-658-38212-4

Printed in the United States
by Baker & Taylor Publisher Services